# Protocol Politics

**Information Revolution and Global Politics**
William J. Drake and Ernest J. Wilson III, editors

# Protocol Politics

## The Globalization of Internet Governance

Laura DeNardis

The MIT Press
Cambridge, Massachusetts
London, England

For information about special quantity discounts, please email special_sales@ mitpress.mit.edu.

This book was set in Stone Sans and Stone Serif by SNP Best-set Typesetter Ltd., Hong Kong. Printed and bound in the United States of America.

Library of Congress Cataloging-in-Publication Data

DeNardis, Laura, 1966–
Protocol politics : the globalization of Internet governance / Laura DeNardis.
    p.  cm.—(Information revolution and global politics)
Includes bibliographical references and index.
ISBN 978-0-262-04257-4 (hardcover : alk. paper)  1. Internet governance.  2. Tele-communication—International cooperation.  I. Title.
TK5105.8854.D46 2009
384.3′3–dc22

                                                                          2008044249

10 9 8 7 6 5 4 3 2 1

*For Deborah R. Smith*

# Contents

# Acknowledgments

*Protocol Politics* follows directly from my doctoral dissertation. Janet Abbate, mentor and friend, provided invaluable contributions to this work. The genesis of this project was a research paper written for Janet's Culture, Politics, and History of the Internet course. In the following years, countless discussions with Janet over coffee crystallized the theoretical and thematic directions of this project. Scott Hauger provided intellectual guidance, encouragement, and ideas that greatly influenced this project as well as many other endeavors. I am also grateful to Barbara Allen for her significant theoretical influence and ongoing support and to Gary Downey, Saul Halfon, and Richard Hirsh for important contributions.

Many of the ideas in this book gelled in the halls and classrooms of Yale Law School and in my intellectual home, the Yale Information Society Project. In particular, I wish to thank Jack Balkin for his extraordinary influence on my still evolving thinking about Internet protocols and Internet governance. I also wish to acknowledge the influence of others in the Yale Information Society Project community, including Anupam Chander, Susan Crawford, Shay David, Laura Forlano, Mike Godwin, James Grimmelmann, Eddan Katz, Christopher Mason, Beth Noveck, Katherine McDaniel, Lea Shaver, Priscilla Smith, Julia Sonnevend, Elizabeth Stark, Madhavi Sunder, Hong Xue, Michael Zimmer, and many others. Thanks also to the students in my Access to Knowledge course for their great ideas.

I also wish to express my gratitude to William Drake and Ernest Wilson, editors of the MIT Press Information Revolution and Global Politics series. A special thanks for the support and contributions of Marguerite Avery and Dana Andrus at the MIT Press and the constructive suggestions of three anonymous reviewers.

I wish to acknowledge the ongoing influence of Milton Mueller on my work and am grateful for finding an intellectual community in the Global Internet Governance Academic Network. I also appreciate conversations with Vinton Cerf and others in the Internet governance community, and I especially benefited from the rich historical archive of documents, standards, minutes, and discussion boards that the Internet Engineering Task Force and the Internet Architecture Board have openly and freely published.

Finally, I thank my family and friends for their love and encouragement, especially Deborah Smith, Robyn Slater, Leonard DeNardis Sr., Gail DeNardis, Leonard DeNardis Jr., Denise DeNardis, Marissa DeNardis, Abby DeNardis, Natalie DeNardis, David Smith, and Elva Smith.

# Protocol Politics

# 1 Scarcity and Internet Governance

Will we shoot virtually at each other over the Internet? Probably not. On the other hand, there may be wars fought about the Internet.[1]
—Vinton Cerf

The Internet is approaching a critical point. The world is running out of Internet addresses. A tacit assumption of the twenty-first century is that sustained Internet growth will accompany the contemporary forces of economic and technological globalization. The ongoing global spread of culture and ideas on the Internet is expected to promote economic opportunity, human flourishing, and the ongoing decentralization of innovation and information production. This possibility is not preordained. It requires the ongoing availability of a technology commons in which the resources necessary for exchanging knowledge are openly and abundantly available. It depends on the availability of open technical protocols on which technological universality and the pace of innovation and access is predicated. It also requires Internet governance frameworks reflecting principles of openness and equal participation.

## Scarcity

At the level of technical architecture, the success and growth of the global Internet is straining critical Internet resources, protocol arrangements, and Internet governance structures. Internet Protocol (IP) addresses are one of the resources necessary for the Internet's ongoing global expansion. Each device that exchanges information over the Internet possesses a unique numerical

---

1. Quote from TCP/IP creator Vinton Cerf in "What I've Learned: Vint Cerf," in *Esquire*, April 2008. Accessed at http://www.esquire.com/features/what-ive-learned/vint-cerf-0508.

address identifying its virtual location, somewhat analogous to a unique postal address identifying a home's physical location. This number is assigned either permanently to a computing device or temporarily for an Internet session. Information is broken into small units, called packets, before routed to its destination over the Internet. Each packet contains the Internet address for both the transmitting device and the receiving device and routers use these addresses to forward packets to their appropriate destinations.

Internet addresses are not an infinite resource. Approximately 4.3 billion available addresses serve the Internet's prevailing technical architecture. These finite resources are not material or natural resources like oil reserves, clean air, or the food supply; they exist at a much more invisible and deeper level of abstraction. They are the critical resources necessary for fueling the global knowledge economy. The traditional technical standard for Internet addresses, called IPv4 or Internet Protocol version 4, originated in the early 1980s and specifies a unique 32-bit number—a series of 32 0s and 1s such as 01101001001010100101100011111010—for each Internet address.[2] This binary number is read by computers, but humans usually express Internet addresses using a shorthand notation called "dotted decimal format" expressed as four octets such as 20.235.0.54.

The address length of 32 bits provides a theoretical reserve of $2^{32}$, or approximately 4.3 billion unique Internet addresses. Internet engineers determined the size of the pool of Internet addresses, usually called the Internet address space, in an era prior to the widespread proliferation of home computers and a decade before the development of the World Wide Web. Establishing a reserve of billions of Internet addresses in this context seemed almost profligate and, in retrospect, demonstrated enormous foresight and optimism about the Internet's future.

But in the twenty-first century, 4.3 billion seems insufficient to meet the demands of projected Internet growth and emerging applications. In 2008 an estimated 1.5 billion individuals used the Internet, a usage rate of, at most, 25 percent of the world's six to seven billion inhabitants. At that same time only 17 percent of the 4.3 billion Internet addresses were still available,[3] with an assignment rate of approximately 160 million per

2. Jon Postel, "DOD Standard Internet Protocol," RFC 760, January 1980. This RFC documents the original Internet Protocol specification. See also Jon Postel, "Internet Protocol, DARPA Internet Program Protocol Specification Prepared for the Defense Advanced Research Projects Agency," RFC 791, September 1981.

3. The allocation of the IPv4 address space is consistently documented on the website of the Internet Assigned Numbers Authority (IANA), the institution

year.[4] Newer Internet applications such as Voice over Internet Protocol (VoIP), Internet television, networked appliances, and mobile Internet devices have only begun to place demands on Internet addresses. Internet engineers forecasted that this pace of innovation and growth would completely exhaust the remaining Internet addresses sometime between 2011 and 2015.

The Internet standards community identified the potential depletion of these 4.3 billion addresses as a crucial technical design concern in 1990. At the time the Internet was primarily an American endeavor and US institutions had already received substantial IP address assignments. As the Internet began to expand internationally, Internet engineers expressed concern that the remaining address reserve would not meet mounting access demands or sufficiently accommodate new technologies such as wireless Internet access and Internet telephony. Even though fewer than 15 million individuals used the Internet in the pre-web technical context of 1990, the Internet standards community anticipated an eventual shortage and began crafting conservation strategies and technological measures to address resource constraints related to IP addresses. Short-term measures such as network address translation (NAT) and classless interdomain routing (CIDR pronounced "cider") have helped postpone somewhat the depletion of the IPv4 address place.

Against the backdrop of competing international protocols and a mixture of political and economic questions, the Internet Engineering Task Force (IETF), the standards-setting institution historically responsible for core Internet protocols, recommended a new protocol, Internet Protocol version 6 (IPv6), to expand the Internet address space. Originally designated the *next generation Internet protocol* (IPng), the IPv6 standard expanded the length of each address from 32 to 128 bits, supplying $2^{128}$, or 340 undecillion unique addresses. The easiest way to describe the multiplier undecillion, at least in the American system, is a 1 followed by 36 zeros.

---

responsible for global coordination of Internet addresses and other number resources. See, for example, "IPv4 Global Unicast Address Assignments." Accessed at http://www.iana.org/assignments/ipv4-address-space.

4. See Internet engineer Geoff Huston's account "IPv6 Deployment: Just Where Are We?" on *Circle ID*, March 2008. Accessed at http://www.circleid.com/posts/ipv6_deployment_where_are_we.

The protocol selected to become the next generation Internet protocol was not the only option and projected address scarcity was not the only concern. The selection was not straightforward. It involved complex technical choices, controversial decisions, competition among information technology companies, resistance from large American companies to the introduction of any new protocols, and an institutional choice between a protocol developed within the prevailing Internet governance institutions and one promoted by a more international institution. Those institutionally involved in Internet standards governance also recognized, in the context of a globally expanding Internet, international concerns about Americans controlling Internet governance functions such as the assignment of IP addresses and the development of core Internet protocols.

Despite the availability of formal IPv6 specifications and its widespread availability in products, and despite the looming depletion of the (IPv4) Internet address space, the upgrade to IPv6 has *barely begun*. The press, technical communities, and IPv6 advocates have forecasted an imminent conversion to IPv6 for more than a decade. Beginning in 2000, governments in Japan, Korea, China, India, and the European Union established national strategies to upgrade to IPv6. These governments have designated the new protocol as a solution to projected address shortages and also as an economic opportunity to develop new products and expertise in an American dominated Internet industry. In contrast to international address scarcity concerns, US corporations, universities, and government agencies have historically possessed ample IP addresses. The United States, with abundant Internet addresses and a large installed base of IPv4 infrastructure, remained relatively dispassionate about IPv6 until discussions commenced in the area of cybersecurity and the war on terrorism after the terrorist attacks of September 11, 2001. The US Department of Defense formally established a directive mandating a transition to IPv6 by 2008, citing a requirement for greater security and demand for more addresses for military combat applications.[5] IPv6 advocacy groups have cited international imbalances in address allocation statistics as indicative of the standard's significance and have described IPv6 as a mechanism for spreading democratic freedoms, promoting economic development, and improving Internet security.

5. US Department of Defense Memorandum issued by DoD chief information officer, John P. Stenbit for Secretaries of the Military Departments, Subject: "Internet Protocol Version 6 (IPv6)," June 9, 2003. Accessed at http://www.dod.gov/news/Jun2003/d20030609nii.pdf.

These government directives and global IPv6 advocacy efforts have not helped spur significant adoption of IPv6. The success of the protocol depends on critical mass of IPv6 deployment, even among those who do not need it. Many market factors have constrained IPv6 adoption, but technical circumstances have also complicated the upgrade. The distributed and decentralized nature of the Internet's technical architecture precludes the possibility of a coordinated and rapid transition. Areas of centralized coordination exist in the development and administration of technical protocols, but decisions about protocol adoption are decentralized and involve the coordinated action of Internet operators and service providers, governments, and individuals overseeing countless network components and segments that comprise the global Internet. The transition, assuming it happens, can only happen incrementally.

More significant, the new protocol is not directly backward compatible with the prevailing protocol in that a computing device exclusively using IPv6 protocols cannot directly exchange information with a computing device exclusively using IPv4. In other words, an individual using an IPv6-only computing device cannot, without some transition mechanism, directly access the majority of web servers that exclusively use IPv4. The transition usually involves the incremental step of deploying both IPv4 and IPv6 protocol suites or implanting one of several technical translation intermediaries. Most upgrades to IPv6 involve dual protocol stack implementations using both IPv4 and IPv6. Projected scarcity in the IPv4 address space was the original incentive for introducing the new protocol, so IPv6 upgrade strategies that also require IPv4 addresses defeat this purpose. The incentive structure for upgrading to IPv6 is paradoxical. Those wanting (or needing) to implement IPv6 have an incentive to do so but are somewhat dependent on IPv4 users adding IPv6 functionality. The incentive for IPv4 users to add IPv6 functionality is for "the common good" rather than for immediate gain.

The Internet Protocol is only one of thousands of information technology standards, but it is the central protocol required in nearly every instance of Internet use. Computing devices that use IP are on the "Net." IPv6 is a critical issue because it was designed to address the problem of projected Internet address scarcity in the context of globalization. It also serves as a useful case study for how protocols, while often established primarily by private actors, are intertwined with socioeconomic and political order. *Protocol Politics* examines what is at stake politically, economically, and technically in the development and adoption of Internet protocols and the scarce resources they create. It explores the implications of looming Internet address scarcity and of the slow deployment of the new protocol designed to address this problem.

## Protocols

A central thesis of this book is that protocols are political. They control the global flow of information and make decisions that influence access to knowledge, civil liberties online, innovation policy, national economic competitiveness, national security, and which technology companies will succeed. From a technical standpoint, protocols can be difficult to grasp because they are intangible and often invisible to Internet users. They are not software code nor material products but are language—textual and numerical language. They are the blueprints that enable technical interoperability among heterogeneous technology products. Technical protocols are functionally similar to real-world protocols. Cultural protocols are not necessarily enshrined in law, but they nevertheless regulate human behavior. In various cultures, protocols dictate how humans greet each other, whether shaking hands, bowing, or kissing. Protocols provide rules for communicating through language with a shared alphabet and grammatical approach, and conventions for mailing a letter. The information content on an envelope bears the recipient's name and address in a predetermined format. There is nothing preordained about these communications norms. They are socially constructed protocols that vary from culture to culture. Instead of providing order to real-world language and human interaction, technical protocols provide order to the binary streams (0s and 1s) that represent information and that digital computing devices use to specify common data formats, interfaces, networking conventions, and procedures for enabling interoperability among devices that adhere to these protocols, regardless of geographical location or manufacturer.

As a note on terminology, this book will use the term "protocol" synonymously with the term "technical standard," although protocol is often a subset of technical standards referring primarily to networking standards that control and enable the flow of information between computing devices on a network as opposed to other types of technical standards such as data file formats or application-level standards.

Understanding the social implications of Internet protocols requires some understanding of which standards fall within this "Internet protocols" taxonomy as well as the Internet governance processes that control these protocols. Most Internet users are familiar with well-known standards such as Bluetooth wireless, Wi-Fi,[6] the MP3[7] format for encoding and compressing audio files, and HTTP,[8] which enables the standard exchange of information between web browsers and web servers. These are only a few examples of thousands of standards enabling the production, exchange, and use of information.

The Internet is based on a common protocological language. The fundamental collection of protocols on which the Internet operates is TCP/IP. By its strict nomenclature, TCP/IP is actually two protocols: Transmission Control Protocol (TCP) and Internet Protocol (IP). In Internet vernacular, however, the term TCP/IP has a more taxonomical function of encompassing a large family of protocols, historically including protocols for electronic mail such as Simple Mail Transport Protocol (SMTP); for file transfer including File Transfer Protocol (FTP); an assortment of routing protocols; and protocols for information exchange between a web client and web server such as HTTP. IPv4 and IPv6 are two fundamental Internet protocols considered components of TCP/IP.

The TCP/IP suite traditionally groups protocols into four functional layers: the Link layer, the Internet layer, the Transport layer, and the Application layer. The Link layer refers to protocols defining the interfaces between a computing device and a transmission medium and is closely associated with local area network (LAN) standards such as Ethernet. The Internet layer includes standards for network-layer addressing and for how packets are routed and switched through a network. The most prominent example of a standard operating conceptually at this level is the Internet Protocol, including both IPv4 and IPv6. Two important examples of Transport-layer protocols are TCP and User Datagram Protocol (UDP), standards responsible for ensuring that information has successfully been exchanged between two network nodes. Finally, the Application-layer protocols interact with actual applications running on a computer and include critical Internet protocols such as HTTP for web communications and FTP for exchanging files. Figure 1.1 depicts a handful of representative protocols traditionally considered part of the TCP/IP family of protocols.

The Internet's core TCP/IP protocols represent only a portion of the standards required for end-to-end interoperability over the Internet. The Internet's routine support of audio, images, and video has expanded the number of embedded standards necessary for any exchange of information over the Internet. Efficient and universal Internet use requires file format and compression

6. The IEEE 802.11 wireless LAN standards are collectively referred to as "Wi-Fi."
7. MPEG Audio Layer 3.
8. HyperText Transfer Protocol.

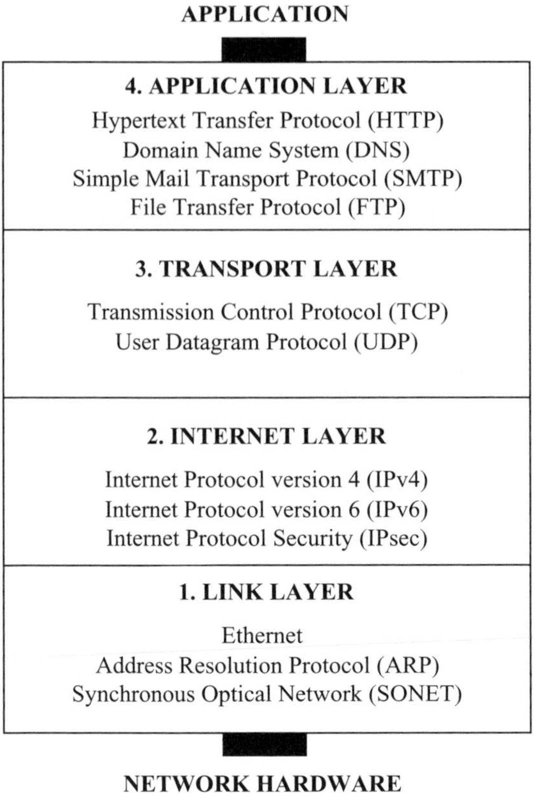

APPLICATION

| **4. APPLICATION LAYER** |
| Hypertext Transfer Protocol (HTTP) |
| Domain Name System (DNS) |
| Simple Mail Transport Protocol (SMTP) |
| File Transfer Protocol (FTP) |
| **3. TRANSPORT LAYER** |
| Transmission Control Protocol (TCP) |
| User Datagram Protocol (UDP) |
| **2. INTERNET LAYER** |
| Internet Protocol version 4 (IPv4) |
| Internet Protocol version 6 (IPv6) |
| Internet Protocol Security (IPsec) |
| **1. LINK LAYER** |
| Ethernet |
| Address Resolution Protocol (ARP) |
| Synchronous Optical Network (SONET) |

NETWORK HARDWARE

**Figure 1.1**
Traditional TCP/IP protocol suite

standards such as MP3 for audio files, JPEG for image files, and MPEG for video. VoIP is another critical area of standardization including prominent protocols such as H.323, Real-time Transport Protocol (RTP), and Session Initiation Protocol (SIP). The types of devices accessing the Internet are equally heterogeneous and include cell phones and other handheld devices, household appliances, and laptops. Internet access standards such as the Wi-Fi family of protocols for wireless laptop connectivity, Bluetooth, or GSM for cell phone connectivity are protocols required for routine Internet use.

Private, non–state institutions and some public–private institutions are responsible for the bulk of Internet standards development. The IETF has developed the majority of Internet standards. As an institution it is unincorporated, has no formal membership or membership requirements, and makes decisions based on rough consensus. The IETF, as the developer of the original Internet Protocol and IPv6, will figure prominently in this book. The World Wide Web Consortium (W3C) is an important, non–state entity that sets Application-layer standards for the web. The International Telecommunications Union's Telecommunications Sector (ITU-T) sets Internet-related standards in areas such as voice over the Internet and security. ITU-T recommendations require consensus and approval of member states. The IEEE (the Institute of Electrical and Electronics Engineers) is a nonprofit professional organization that has contributed many key networking standards ranging from various incarnations of the Ethernet LAN standard to the Wi-Fi family of standards. These are only a few of many institutions involved in Internet standards governance.

This book focuses most heavily on the Internet Protocol. IP has several characteristics that place it at the center of a number of social, economic, and institutional concerns. The first quality is *universality*—IP is a necessary precondition to being on the Internet. Nearly every information exchange over the Internet uses IP. Referring back to Figure 1.1, it is notable that at three of the four protocol levels, there are protocol alternatives. The Transport-layer function can easily include UDP or TCP; any number of LAN technologies can achieve Link-layer functionality; the protocol used at the Application layer is dependent on the application in question (e.g., email, web, voice). At the Internet layer, the primary protocol is IP. Whether IPv4 or IPv6 is being used, IP is the defining protocol for network level functionality. If IP is the least common denominator for communicating over the Internet and the one protocol used in every instance of Internet connectivity, one can envision that this protocol would be relevant to a number of concerns and of interest to those seeking greater control of the Internet.

A second characteristic of IP is *identification*—IP creates a globally unique identifier. As the Internet architecture is currently constituted, no two computing devices can simultaneously use the same address. Regardless of whether an IP address is permanently assigned to a computing device or assigned temporarily for a session, the IP address, along with other information, can potentially provide information about what computing device conducted a specific activity on the Internet at a specific moment in time.

A third characteristic of IP is *exposure*—IP addresses are not encrypted. An important design consideration that potentially factors into concerns about privacy, censorship, and access is that IP addresses are usually "out in the open" on the Internet. Even when information is encrypted for transmission over the Internet, the packet header appended to this information is not necessarily encrypted. IP addresses are included in this header. Given that IP addresses are not encrypted, it is always conceivable to determine the IP address attached to content, even if the content itself is cryptographically protected.

A fourth characteristic is *disinterestedness*—IP locates intelligence at end points. Although this principle is not exclusive to IP, a traditional design feature underlying Internet protocols is to locate intelligence at network end points. Applying this principle to IP, this protocol would not be concerned with the content of packets transmitted over the Internet, or whether the content was viewed, but only with the efficient routing and addressing necessary for the packet to reach its end point.

Examining Internet standardization and the Internet Protocol is an inherently interdisciplinary exercise involving technology, culture, politics, institutional economics, and law. To confront this inherent interdisciplinarity, *Protocol Politics* is heavily influenced by the field of Science and Technology Studies (STS); accounts of standards as political from Janet Abbate and other historians of technology; the work of legal scholars such as Jack Balkin, Yochai Benkler, Larry Lessig, Anupam Chander, and Madhavi Sunder; and the field of institutional economics, particularly as applied by Internet governance scholar, Milton Mueller.

Politics are not external to technical architecture. As sites of control over technology, the decisions embedded within protocols embed values and reflect the socioeconomic and political interests of protocol developers. In a discussion about debates over Open Systems Interconnection (OSI) versus TCP/IP in *Inventing the Internet*, Janet Abbate notes that technical standards are often construed as neutral and therefore not historically interesting. Perceptions of neutrality derive in part from the esoteric and concealed nature of network protocols within the broader realm of information technology. As Abbate demonstrates, "The debate over network protocols illustrates how standards can be politics by other means. . . . Efforts to create formal standards bring system builders' private technical decisions into the public realm; in this way, standards battles can bring to light unspoken assumptions and conflicts of interest. The very passion with which stakeholders contest standards decisions should alert us to the deeper meanings beneath the nuts and bolts."[9] Many of the research questions *Protocol Politics* examines emanate from Abbate's view about debates over protocols bringing to light unspoken conflicts of interest.[10]

9. Janet Abbate, *Inventing the Internet*, Cambridge: MIT Press, 1999, p. 179.
10. Like Abbate's account, other historical works similarly reinforce this political dimension of technical standardization. For example, Ken Alder's account of the development of the metric standard during the French Revolution, *The Measure of All Things: The Seven-Year Odyssey and Hidden Error That Transformed the World* (New York: Free Press, 2002), examines how seemingly neutral and objective standards are historically contingent and embody both political and economic interests.

*Protocol Politics* also asks questions about how protocols, once developed, have political meanings that can be adapted for various purposes.[11] The decisions made during protocol design can have significant public policy consequences. From an advocacy standpoint, the Internet Standards, Technology and Policy Project at the Center for Democracy and Technology (CDT) in Washington, DC, has raised awareness about the public policy consequences of Internet standards. Increasingly, policy decisions about whether to advance or restrict online freedoms occur in the technical standardization process invisible to the public and established primarily by private industry rather than legislatures. When Internet engineers designed the Internet address structure for the new IPv6 standard, they decided to build some privacy protections into the protocol. The CDT's project sought to increase public awareness and to inject a public voice into this technology-embedded form of public policy.[12]

Standards are not software code but language. If code is "law"[13] regulating conduct similar to legal code, or even if software is its own modality of regulation unlike law or physical architecture,[14] then the underlying protocols to which software and hardware design conforms represent a more embedded and more invisible form of legal architecture able to constrain behavior, establish public policy, or restrict or expand online liberty. In this sense, protocols have political agency—not a disembodied agency but one derived from protocol designers and implementers. There is no remote corner of the Internet not dependent on protocols. They are control points, in some cases, areas of centralized control, and sometimes distributed control, mediating tensions between order and freedom.

11. See, for example, Paul Edwards's critical integration of political and technical histories in *The Closed World, Computers and the Politics of Discourse in Cold War America* (Cambridge: MIT Press, 1996), examining how cold war "politics became embedded in the machines—even, at times, in their technical design—while the machines helped make possible its politics." (p. ix).

12. See, for example, Standards Bulletin 2.01, "ENUM and Voice over Internet Technology," April 28, 2003; Standards Bulletin 1.03, "Patents on Internet Technology Standards," December 13, 2002; John Morris and Alan Davidson, "Policy Impact Assessments: Considering the Public Interest in Internet Standards Development," 2003; and Alan Davidson, John Morris, and Robert Courtney, "Strangers in a Strange Land: Public Interest Advocacy and Internet Standards," 2002. Papers accessed at http://www.cdt.org/standards.

13. Lawrence Lessig, *Code and Other Laws of Cyberspace*, New York: Basic Books, 1999.

14. James Grimmelmann, "Regulation by Software," 114 *Yale Law Journal* 1719 (2005).

Internet protocols are an example of what Yochai Benkler calls *knowledge-embedded tools*, similar to enabling technologies for medical and agricultural resources.[15] Knowledge-embedded tools, such as open (vs. proprietary) standards, are necessary for enhancing welfare and enabling innovation itself. Internet standards such as TCP/IP and HyperText Markup Language (HTML) have historically been openly available, enabling citizens and entrepreneurs to contribute to Internet innovation, culture, and electronic discursive spheres. Other widely used technical standards do not exhibit this same degree of openness. From an economic standpoint, standards have significant effects such as enabling or restricting global trade and enabling competition and innovation in product areas based on common standards. [16] As David Grewal suggests in *Network Power*, the "creation and diffusion of standards underlying new technologies is a driving element of contemporary globalization."[17]

A striking feature of this type of social force is that it is established by institutions, often private institutions, rather than by elected representatives. Following Milton Mueller's approach in *Ruling the Root: Internet Governance and the Taming of Cyberspace*, this book draws from institutional economics—the intersection of law, economics, and politics. Much work has been done on the critical role of institutions in creating the world around us.[18] *Protocol Politics* examines institutional dynamics but also highlights the critical contributions of key individuals in the evolution of Internet governance and their contributions to the rise of new production models embraced by Internet governance institutions. These models transcend national boundaries, bypass intergovernmental organizations, and challenge traditional beliefs about economic behavior. One objective of this book is to examine the institutional

15. Yochai Benkler, *The Wealth of Networks: How Social Production Transforms Markets and Freedom*, New Haven: Yale University Press, 2006.
16. See Rishab Ghosh, *An Economic Basis for Open Standards*, December 2005. Accessed at http://flosspols.org/deliverables/FLOSSPOLS-D04-openstandards-v6.pdf.
17. David Grewal, *Network Power: The Social Dynamics of Globalization*, New Haven: Yale University Press, 2008, p. 194.
18. For example, Arturo Escobar suggests, "The work of institutions is one of the most powerful forces in the creation of the world in which we live," in *Encountering Development, The Making and Unmaking of the Third World*, Princeton: Princeton University Press, 1995, p. 107. See also Yochai Benkler, "Coase's Penguin, or, Linux and the Nature of the Firm," 112 *Yale Law Journal* 369 (2002), for an exploration of new "commons-based peer-production" models of large-scale collaboration motivated by a variety of incentives distinct from managerial hierarchy or market prices.

characteristics and principles necessary to maximize the legitimacy of private institutions to establish global knowledge policy.

## An Internet Governance Framework

Questions about Internet standardization and the IP address space are questions about Internet governance. While the distributed architecture and ubiquity of the Internet can convey the impression that no one controls the Internet, coordination—sometimes centralized coordination—occurs in several technical and administrative areas necessary to keep the Internet operational. John Perry Barlow, in *A Declaration of the Independence of Cyberspace* written to traditional world governments, wrote that "We are forming our own Social Contract. This governance will arise according to the conditions of our world, not yours. Our world is different."[19] But there have always been some centralized governance functions in cyberspace, although not governance by sovereign governments or even intergovernmental organizations.

The term "Internet governance" has many definitions and is a highly contested term.[20] Internet governance functions have been around for far longer than the term Internet governance. Even the term "governance" in this context requires qualification because Internet governance actors have not primarily been governments. As Milton Mueller explains, there are sometimes two extreme views about who controls the Internet: the view that the Internet is inherently uncontrollable and therefore not controlled; and the antithetical view that a small cabal of individuals and corporations has authoritative hegemony over the Internet. As Mueller suggests, "For any complex sociotechnical system, especially one that touches as many people as the Internet, control takes the form of *institutions*, not commands."[21] The functions these institutions control can be quite expansive, depending on how one defines Internet governance.

19. John Perry Barlow, "A Declaration of the Independence of Cyberspace," 1996. Accessed at http://homes.eff.org/~barlow/Declaration-Final.html.
20. See Jeanette Hoffman, "Internet Governance: A Regulatory Idea in Flux," 2005. English translation accessed at http://duplox.wzb.eu/people/jeanette/texte/Internet%20Governance%20english%20version.pdf.
21. Milton Mueller, *Ruling the Root*, Cambridge: MIT Press, 2002, p. 11.

Internet governance refers generally to policy and technical coordination issues related to the exchange of information over the Internet. Many conceptions of Internet governance, especially those emanating from technical communities, are quite bounded in scope, describing Internet governance as having three distinct functions: "(1) technical standardization, (2) resource allocation and assignment, and (3) policy formulation, policy enforcement, and dispute resolution."[22] Many Internet governance examinations inquire within a closed sphere of institutional interactions and their internal technical decision-making processes. This type of inquiry does not necessarily reflect the contextual milieu that shapes decisions or the broader social implications of these decisions. The underlying framework of *Protocol Politics* rests on a broader view of Internet governance to create openings for examining how values shape Internet governance decisions and for assessing the economic, legal, and political externalities of these decisions.

In addition to Internet standardization there are four additional areas of Internet governance, with Internet governance broadly conceived: critical Internet resources, intellectual property rights, security, and communication rights.

### Critical Internet Resources

In regard to critical Internet resources, the topic that receives the most press and scholarly attention is the role of ICANN as a global governance institution and its associated policies about the management and assignment of Internet domain names and numbers. Most of this concern addresses domain names. The domain name system (DNS) serves a critical function necessary for the successful operation of the Internet, translating between alphanumeric domain names and their associated numerical IP addresses necessary for routing information across the Internet. The DNS performs this address resolution process and resolves billions of queries each day. The DNS is really an enormous database management system distributed globally across numerous servers and operating like a hierarchical tree. The component (.gov, .edu, .com, etc.) on the far right of any domain name is called the top-level domain (TLD). Other top-level domains are country codes, or ccTLDs, such as .br for

22. Internet Governance Project White Paper, "Internet Governance: The State of Play," September 2004. Accessed at http://www.internetgovernance.org/pdf/ig-sop-final.pdf. The Internet Governance Project is a partnership of scholars at Syracuse University, Georgia Institute of Technology, and Wissenschaftszentrum Berlin für Sozialforschung.

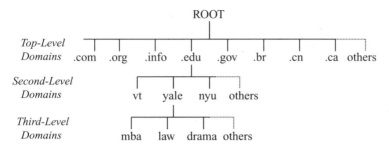

**Figure 1.2**
Domain name space

Brazil, .ca for Canada, and .cn for China. In domain name semantics, the word to the left of the top-level domain is called the second-level domain, such as the "yale" in "yale.edu." Figure 1.2 conceptually depicts a small portion of the domain name space. The Internet's root name servers contain a master file known as the root zone file itemizing the IP addresses and associated names of the official DNS servers for all top-level domains.

The domain name system establishes the domain name space in the same way that the Internet Protocol establishes the Internet address space. As critical resources necessary for Internet connectivity and use, the management of the Internet address space and the domain name space are central tasks of Internet governance. This function includes the actual allocation and global coordination of Internet domain names and numbers. Within ICANN, the Internet Assigned Numbers Authority (IANA) is responsible for root zone management for the DNS, as well as globally coordinating the IP address space. Internet governance concerns about the DNS include controversies about the assignment of top-level domain names, conflict over authority and control over the root zone file and root name servers, issues of national and transnational jurisdiction, questions about institutional legitimacy, and a host of policy questions dealing with critical infrastructure protection, intellectual property issues related to domain names, dispute resolution, and institutional questions of legal and political responsibility.

One objective of *Protocol Politics* is to bring more attention to the IP address space in the Internet governance realm of critical Internet resource management. A major analytical theme will address how new technologies create new resources. This theme is not unique to Internet governance. Battles over technologically derived resources are a central issue of information and communication technology policy, whether addressing electromagnetic spectrum or bandwidth in network neutrality debates. What may be unique about Internet addresses is that they are a completely global resource that has always been centrally coordinated by some Internet governance entity. The Internet Protocol (both IPv4 and IPv6) created Internet addresses. In the case of the prevailing IPv4 protocol, the resource pool contains a theoretical maximum of approximately 4.3 billion addresses. The IPv6 address space contains 340 undecillion addresses. Like electromagnetic spectrum and other technologically derived resources, Internet addresses carry significant network externalities and economic value. This value cannot be assessed within the traditional sphere of market economics because, as of yet, these finite resources have never been exchanged in free markets. Centralized control of IP addresses has historically existed to maintain the architectural principle of globally unique addresses. A single individual, Jon Postel, originally administered these finite technical resources but responsibility gradually shifted to geographically distributed, international registries known as regional Internet registries (RIRs). Despite this global dispersion of IP addresses and assignment responsibility, definitive oversight of the entire address reserve, including the allocation of address resources to international registries, has remained centralized, eventually becoming an IANA administrative function under ICANN.

The extent to which Internet addresses have critical technical, economic, and political implications raises governance questions about how access to resources and power over these resources are distributed or should be distributed among institutions, nation-states, cultures, regions, and among entities with a vested economic interest in the possession or control of these resources. This book examines IP address creation and distribution not only from the standpoint of institutional economics and efficiency, but from normative and overarching questions of distributive justice.[23]

### Intellectual Property Rights

In addition to critical resource management, intellectual property rights are a significant Internet governance concern. Decisions related to intellectual property rights order the flow of information, creativity, and compensation over the Internet. This area encompasses issues such as trademarks, patents, and copyright, and the balance between intellectual

---

23. Anupam Chander explains that, in cyberlaw scholarship generally, concerns about human values such as distributive justice and equality are greatly neglected. See Anupam Chander, "The New, New Property," 81 *Texas Law Review* 715 (2003).

property protection and the Internet's tradition of free and open access to knowledge. One objection to including intellectual property as an Internet governance concern is the argument that Internet governance should only address technical architecture and critical resources, not content. This argument quickly breaks down because intellectual property rights enforcement is often implemented within technical architecture, such as copyright filtering or digital rights management (DRM) technologies and because some of the greatest intellectual property concerns address technical architecture itself rather than content. Copyright and patents in technical standardization are particularly complex areas intersecting with innovation policy, antitrust concerns, economic competition, and the openness of the Internet. Intellectual property scholar Mark Lemley describes the problem of patent owner holdup, particularly in the technical standardization context, as "the central public policy problem in intellectual property law today."[24]

Intellectual property questions are also at the heart of many domain name controversies, such as trademark disputes over domain name registrations. Traditional legal remedies for Internet trademark disputes have not always been helpful because of uncertainty about which country's laws have jurisdiction in any given dispute and because traditional legal intervention is a lengthy process relative to the pace of Internet developments. ICANN's Uniform Domain-Name Dispute-Resolution Policy (UDRP) has served as a mechanism for trademark protection in the sphere of domain names but, like most of ICANN's activities, has been controversial.

Intellectual property rights for content itself can also be a purview of Internet governance institutions, particularly if one views intellectual property issues as more about social relations and the ability of humans to engage in cultural production and meaning and free expression.[25] A central question is how to view "fair use" in online environments and how to balance the goal of protecting artists' and authors' rights with a separate set of public interest questions such as improving access to knowledge in the developing world, encouraging digital education, and facilitating the creation of culture and the ability to dissent. Online copyright protection not only places restrictions on copying a work similar to restrictions in the offline world, it can mean additional

24. Mark Lemley, "Ten Things to Do about Patent Holdup of Standards (and One Not To)," 48 *Boston College Law* 149 (2007).
25. See, generally, Madhavi Sunder, "IP3," 59 *Stanford Law Review* 257–332 (2006). "Intellectual property is about social relations and should serve human values."

restrictions in access through technological and legal measures for copyright protection.

Internet governance questions addressing intellectual property occur at many levels. For companies providing Internet services based on common technical standards, one concern is whether they are liable if they host copyright-infringing content. Institutionally, standards-setting organizations sometimes have intellectual property policies such as requiring ex ante disclosure of intellectual property rights among member companies involved in standardization or requiring agreements that any standards-based intellectual property rights be made available on a so-called reasonable and nondiscriminatory basis. As mentioned, ICANN has procedures to deal with trademark protection. Other intellectual property related Internet governance takes place at the national level, such as through the Digital Millennium Copyright Act (DMCA) passed in the United States in 1998, and at the international level through the World Intellectual Property Organization (WIPO) or the World Trade Organization's (WTO's) TRIPS agreement, short for Trade-Related Aspects of Intellectual Property Rights.

**Security**
Internet security is perhaps the most critical area of Internet governance. When a worm or denial of service attack compromises the Internet's reliability and availability, all other areas of Internet governance seem irrelevant. This Internet governance is particularly complex because security problems involve a wide variety of concerns ranging from critical infrastructure protection to user authentication and because responsibility for Internet security is distributed so widely in a complex matrix of public and private control.

The universality and openness of the Internet make it a prime target for attacks, whether for reasons of criminal activity, terrorism, or to advance a political agenda. The most publicly understood security problems are viruses, malicious code embedded in software that inflicts damage when the code is executed, and worms, self-replicating and self-propagating code that exploits weaknesses in protocols and software to inflict harm. These types of attacks can be costly. According to congressional testimony, the "I Love You" virus that spread throughout Asia, Europe, and North America affected 65 percent of North American businesses and infected 10 million computers.[26] Distributed denial of service (DDoS) attacks are an even

26. US House of Representatives, Subcommittee on Technology, Committee on Science Hearing on Computer Viruses, May 10, 2000.

greater threat. These attacks hijack computers, which unknowingly work together to disable a targeted computer by flooding it with requests. The targets of these attacks have included the Internet's root servers, high-profile commercial websites, and government servers.[27] Other types of Internet security concerns include identity and password theft, data interception and modification, and bandwidth piracy. Critical infrastructure protection, whether of physical telecommunications infrastructures or on a critical Internet system such as the DNS, is always a concern. Hackers can use computing systems to disrupt physical infrastructures such as when a disgruntled employee broke into a computer system controlling an Australian sewage treatment plant and released millions of liters of raw sewage into the environment.[28]

A key Internet governance question about security asks what are the appropriate roles of national governments, the private sector, individual users, and technical communities in addressing Internet security. The private sector develops and implements the majority of Internet security measures. Businesses selling products and services online implement voluntary authentication and privacy mechanisms such as public key cryptography to secure electronic commerce. Service providers, business Internet users, and individual users implement their own access control mechanisms such as firewalls. Standards institutions such as the IETF and the IEEE develop security-related protocols.

Governments also have a role. Most national governments enact policies for critical infrastructure protection and cybersecurity. For example, the US Department of Homeland Security operates a Computer Emergency Response Team (CERT) that works in conjunction with private industry to identify security problems and coordinate responses. Detecting and responding to Internet security problems is a complicated area of public–private interaction and also one requiring transnational coordination. There are hundreds of CERTs around the globe, many of which are hybrid public–private institutions. The coordination of information and responses to attacks among these public–private entities is a critical Internet governance concern.

27. For a history of some DDoS and other Internet attacks, see Laura DeNardis, "A History of Internet Security," in *The History of Information Security*, Karl de Leeuw and Jan Bergstra, eds., Amsterdam: Elsevier, 2007.
28. Parliament of the Commonwealth of Australia, Parliamentary Joint Committee on the Australian Crime Commission, Cybercrime, March 2004. Accessed at http://www.aph.gov.au/senate/committee/acc_ctte/completed_inquiries/200204/cybercrime/report/report.pdf.

**Communication Rights**

Finally, Internet governance involves concerns about communication rights, particularly when technical architecture design or policy formulation intersects with the public's civil liberties online. Freedom of expression and association are increasingly exercised online and institutional decisions about technical architecture can determine the extent of these freedoms as well as the degree to which online interactions protect individual privacy and reputation. The same technologies that expand freedom of expression have created unprecedented privacy concerns, and Internet governance decisions often must mediate between the conflicting values of free expression and privacy. To the extent that architectural design and implementation decisions and policies determine communication rights, this area should be construed as an important part of Internet governance.

Traditional governments have not historically had the most prominent role in Internet governance, but many communication rights areas that governments have traditionally overseen have converged with Internet infrastructure, raising questions about public versus private Internet control. For example, video delivery no longer depends on traditional broadcast structures, and voice delivery no longer depends on traditional telephone systems. Voice and video have become just like any other application on the Internet, enabled in part by new protocols such as VoIP and Internet Protocol Television. These advancements have complicated Internet governance because of the incompatibilities between prevailing approaches to Internet governance and the governance of traditional media and broadcast. Traditional Internet governance has involved private–public and multistakeholder coordination, has been international in scope, and has embraced the philosophy of making information accessible to everyone. Governments have historically provided traditional broadcast and media oversight. These approaches have been national or regional in scope and have promoted highly controlled flows of information to protect intellectual property and businesses models. Governance models in the context of this convergence are an emerging Internet governance concern, especially to those opposed to the possibility of an increasing role for governments in Internet regulation.

**Organization of *Protocol Politics***

The previous section laid out a broad view of Internet governance. The development of IPv6, on its surface, would seem to involve only two facets of Internet governance: Internet standardization and critical Internet resources. A central theme of this book is that Internet protocols and Internet resource management are not merely issues of establishing technical specifications or administering resources but are issues that traverse all Internet governance concerns sketched out in the framework described above. Protocols involve questions of technical interoperability and the establishment of critical Internet resources, but also questions about intellectual property, security, and communication rights. Many such questions have been traditionally overseen by governments, but they are increasingly being addressed in the technical architecture.

The remainder of *Protocol Politics* is divided into five sections. Chapter 2 examines how protocol selection is a political process as well as a technical issue. The chapter explores how concerns about resource scarcity emerged within the context of Internet globalization, what the alternatives were to IPv6, why they were discarded, and what was at stake in the selection process. The technical standard that became IPv6 was not the only alternative. The Internet engineers selecting the new protocol established a guideline that only technical factors would enter the selection process, but this chapter describes how a significant factor in the selection process appears to have been the selection of which standards-setting institution would have control over Internet standards.

Participants in the Internet standards process first articulated concerns about the Internet running out of addresses in the early 1990s. At the time a set of protocols known as OSI protocols were in competition with Internet protocols to become the universal standard for interconnecting diverse computing environments. The chapter describes how the two final alternatives for the next generation Internet protocol involved a choice between an IETF originating protocol and an OSI-related protocol promoted by the International Organization for Standardization (ISO). If the ISO protocol had been selected, the ability to control and change the key Internet protocol would likely have rested with ISO rather than the IETF, which had historically been responsible for the development of Internet protocols.

By examining IPv6 against its discarded alternatives, this chapter reveals the conflicts among institutions, between trusted insiders and newer participants, and between dominant companies and new entrants, all within the context of increasing Internet globalization. Another chapter theme is the phenomenon of protocol selection occurring extraneous to contemporary forces of market economics.

Chapter 3 examines how the design of protocols can involve decisions that affect the public's civil liberties online. The public policy embedded in

technical standards can present an opportunity either to advance the libertarian ideals historically associated with the Internet's underlying protocols
or to restrict access, regulate speech, or impose censorship. Protocol design
reflects the values of protocol designers. As Internet engineers designed the
technical specifications of IPv6 in the years following its selection, they
weighed design decisions related to issues of Internet user anonymity and
location privacy. The chapter explains the privacy issue that Internet engineers addressed, describes the process whereby Internet engineers opted to
design some privacy protections into the protocol, and recounts contemporaneous concerns raised by privacy advocates, particularly in the European
Union. The chapter addresses the implications of private standards-setting
institutions establishing public policy, the question of institutional legitimacy, and the issue of how, considering technical barriers to public participation, the public interest can realistically enter these decisions.

Chapter 4 examines the politics of protocol adoption, including the
ambitious national IPv6 strategies of governments in China, Japan, the
European Union, Korea, and India. Many of the rationales for upgrading
had less to do with the increasing reality of Internet address depletion than
with promoting other socioeconomic objectives. This chapter suggests
that the *promise* of IPv6 aligned with broader political objectives such as
European unification goals or attempts to reverse economic stagnation in
Asia. The chapter also describes how US politicians began linking the prospect of product development and expertise in IPv6 with the objectives of
fighting a more distributed war on terrorism and improving US economic
competitiveness in the context of globalization and the outsourcing of
American jobs to China and India. The chapter examines how IPv6 advocates and stakeholders also linked the protocol with a number of social
and economic development objectives ranging from global democratic
reform to third world development. One related issue is the role of open
intellectual property rights in Internet standards in opening the possibility
of global competition and innovation. Another is the ongoing narrative
among advocates of IPv6 providing inherently greater security, a promise
that has proved to be highly contestable. Another theme of chapter 4 is
how many governments have rejected laissez-faire protocol adoption in
favor of sweeping government mandates backed by economic inducements. Finally, the chapter describes the most interesting aspect of government IPv6 adoption policies. National protocol upgrade deadlines have
passed with no significant deployment of IPv6. The chapter describes the
transition challenges that have hindered IPv6 implementation and assesses
prospects for the emergence of a transition strategy.

Chapter 5 examines the Internet address space and how technical protocols create new scarce resources. When the value of these resources becomes clear, their possession and control become a source of global tension. The management and control of Internet addresses is a fascinating issue because a centralized actor has always controlled and allocated these resources and because they have never been exchanged in free markets. This chapter examines the origination and allocation of the Internet address space, the emergence of debates about address scarcity, the evolution of control of IP address assignment, and the near depletion of the IPv4 address space. In the context of describing this evolution, the chapter examines three Internet governance questions: (1) the question of who controls (and who should control) the allocation of Internet addresses; (2) the manner in which these scarce resources are allocated, whether directed toward market efficiency, distributive justice, rewarding first movers, or other objective; and (3) the overarching question of whether there exist sufficient addresses to meet current and anticipated demand.

Chapter 6 presents a general framework for understanding the political and economic implications of protocols in their design, implementation, and adoption. Drawing from the history of IPv6 and other protocols, this chapter examines six ways in which technical protocols potentially serve as a form of public policy: (1) the content and material implications of standards can themselves constitute substantive political issues; (2) standards can have implications for other political processes; (3) the selection of standards can reflect institutional power struggles for control over the Internet; (4) standards can have pronounced implications for developing countries; (5) standards can determine how innovation policy, economic competition, and global trade can proceed; and (6) standards sometimes create scarce resources and influence how these resources are globally distributed.

Whereas Internet protocols and other technical standards have broad political and economic implications, issues regarding who decides in matters of standards setting and how they decide are key questions, especially to the extent that private industry engages in the establishment of public policy. The IETF is only one of many organizations setting standards, ranging from physical infrastructure to applications, necessary to enable the universal exchange of information over the Internet. The IETF has a generally open and transparent approach even though many barriers to public participation exist. But other institutions have different standards-setting norms that lack the openness and transparency of IETF processes. This chapter suggests best practices in Internet standards setting based on principles of openness, transparency, and economic competition. The rationale for promoting so-called open standards are technical, economic, and political—with the technical rationale of open standards promoting maximum technical interoperability, the economic rationale of enabling competition and minimizing anticompetitive and monopolistic practices, and the political rationale of maximizing the legitimacy of standards-setting organizations to make decisions that establish public policy in areas such as individual civil liberties, democratic participation, and user choice.

The final section of chapter 6 shifts attention back to IPv6 and the limits of both protocol openness and government intervention in influencing standards adoption. The wide discrepancy between a decade of promises about imminent IPv6 adoption and the reality of slow deployment has been one of the most intriguing stories in the history of the Internet. The chapter concludes by exploring the possible implications of IPv4 address depletion and the slow deployment of IPv6 to global Internet access needs, to Internet governance structures, and to the future of the Internet's underlying architecture.

## 2  Protocol Selection as Power Selection

At the core of "universal standards" commonly taken to be products of objective science lies the historically contingent, and further ... these seemingly "natural" standards express the specific, if paradoxical, agendas of specific social and economic interests.[1]

—Ken Alder, *A Revolution to Measure*

Internet engineers long ago forecasted that Internet addresses would become critically scarce. These concerns surfaced in 1990 in a world in which the web did not yet exist, prior to the founding of Internet companies such as Amazon, Netscape, or Yahoo!, and long before the existence of Google, Facebook, or Wikipedia. The Internet was growing internationally but Americans were still the predominant users and developers. Businesses did not use the Internet to any great extent, and most of the public was unaware of its existence. The Internet's most popular application was text-based email, and it did not yet support voice, video, or images. Fewer than 15 million individuals used the Internet, but the network was expanding internationally. Indeed it was in this latter context that Internet engineers first raised the issue of Internet address scarcity and the need for a new network protocol to increase the number of devices able to connect to the network.[2]

1. Ken Alder, "A Revolution to Measure: The Political Economy of the Metric System in France," in *Values of Precision*, M. Norton Wise, ed., Princeton: Princeton University Press, 1995, pp. 39–71.
2. For example, questions about the possibility of IP address exhaustion were present during an April 26, 1990 Internet Architecture Board meeting, according to the meetings. Accessed at http://www.iab.org/documents/iabmins/IABmins.1990-04-26.html. Similar questions and concerns emerged at the next quarterly IAB meeting, on June 28–29 1990. Accessed at http://www.iab.org/documents/iabmins/IABmins.1990-06-28.html.

## Protocol Globalization

The concern over address scarcity surfaced within an Internet governance institution called the Internet Activities Board (IAB). Understanding the responsibilities of this organization and its relationship to other Internet governance institutions requires recognizing that at the time the IAB had ultimate responsibility for the direction of the Internet's architecture. As outlined by the then–IAB chair, Vinton Cerf, the IAB "(1) sets Internet standards; (2) manages the RFC publication process; (3) reviews the operation of the Internet Engineering Task Force (IETF) and the Internet Research Task Force (IRTF); (4) performs strategic planning for the Internet, identifying long-range problems and opportunities; (5) acts as an international technical policy liaison and representative for the Internet community; and (6) resolves technical issues which cannot be treated within the IETF or IRTF frameworks."[3]

The second function, managing the RFC publication process, refers to the Request for Comments (RFC) series, electronic archives documenting protocols, procedures, and other information related to the Internet's ongoing development since 1969. The thousands of RFCs offer a technical and social history of proposed Internet standards, final Internet standards, and opinions from Internet pioneers. The late Jon Postel served as collector, editor, and archivist of more than 2,500 RFCs for 28 years beginning in 1969. After Postel's death in 1998, his colleague, Joyce Reynolds, assumed these responsibilities, later expanded to a small group of individuals funded by the Internet Society. The entire RFC series is electronically available via www.rfc-editor.org, although the RFCs were originally paper documents, having, as Vinton Cerf described "an almost 19th century character to them—letters exchanged in public debating the merits of various design choices for protocols in the ARPANET."[4] RFCs progress through the standards track categories of "proposed standards," "draft standards," and "standards."[5] Other RFCs, called "best current practices," are not standards but official guidelines issued by the Internet standards community. Some RFCs are "historic," archiving former Internet standards that have been "deprecated," a term describing a standard or information that has become

3. Vinton Cerf, in "The Internet Activities Board," RFC 1120, May 1990, p. 2.
4. RFC Editor, "30 Years of RFCs," RFC 2555, April 7, 1999, p. 4.
5. For a detailed description of the Internet standards review process, see Harald Alvestrand's best current practices document, "The IESG and RFC Editor Documents: Procedures," RFC 3932, October 2004.

obsolete or replaced. Additionally, some RFCs are not standards but are "informational" or "experimental," either originating within or external to the IETF. Several RFCs, often published on April Fools' Day, are actually jokes, such as "Hyper Text Coffee Pot Control Protocol (HTCPCP/1.0)," a lengthy RFC attributing the consumption of the IPv4 address space to the proliferation of networked coffee pots and proposing a new control protocol accordingly.[6]

The IAB had overall responsibility for the Internet because it ultimately approved standards and set the Internet's strategic direction. In 1990 the IAB was made up of eleven individuals, primarily Americans who worked for corporations, universities, and research institutions.[7] Members communicated with each other via electronic mailing lists and also held quarterly meetings to assess the overall condition of the Internet and discuss technical and policy issues. This independent group was closed to general public involvement in that the IAB chair, then Vinton Cerf, appointed members[8] but was open in the sense that it was strongly influenced by the recommendations originating in the open IETF, and that all IAB decisions were made publicly available.[9]

The IAB had been formalized as an institution in 1983, but its origins traced to the late 1970s period of the ARPANET project when researchers involved in protocol development founded an informal committee known as the Internet Configuration Control Board (ICCB). Then DARPA program manager, Cerf, was instrumental in establishing the committee, and David Clark of MIT's Laboratory for Computer Science became the chairman. In 1983, the year TCP/IP became the formal protocol underpinning of the ARPANET, the group renamed the ICCB the Internet Activities Board, or IAB. Vinton Cerf became the IAB's chair in 1989. The organization's primary responsibilities involved oversight of the Internet's protocol architecture and included ultimate responsibility for approving protocols.

6. Larry Masinter, "Hyper Text Coffee Pot Control Protocol (HTCPCP/1.0)," RFC 2324, April 1, 1998.

7. The eleven IAB members in 1990 were Vinton Cerf, chair; Robert Braden (USC-ISI), executive director; David Clark (MIT-LCS), IRTF chair; Phillip Gross (CNRI), IETF chair; Jon Postel (USC-ISI), RFC editor; Hans-Werner Braun (Merit), member; Lyman Chapin (DG), member; Stephen Kent (BBN), member; Anthony Lauck (Digital), member; Barry Leiner (RIACS), member, and Daniel Lynch (Interop, Inc.), member. Source: RFC 1160.

8. Vinton Cerf, "The Internet Activities Board," RFC 1160, May 1990.

9. Ibid.

The IAB had established the Internet Engineering Task Force in 1986 as a subsidiary institution serving as the primary standards organization developing Internet protocol drafts. The IETF has no formal membership, is composed of volunteers, and is a non-incorporated entity with no legal status. The IETF traditionally has held triennial face-to-face plenary meetings. The working climate of these gatherings is informal, with fluid agendas, social gatherings, and a relaxed dress code dominated by "t-shirts, jeans (shorts, if weather permits), and sandals."[10] IETF working groups conduct the bulk of standards development and communicate primarily through electronic mailing lists to which anyone may subscribe. Area directors (AD) head up the working groups and, these ADs (approximately eight at any time) along with the IETF chair constitute the Internet Engineering Steering Group (IESG). Standards percolate up from the IETF working groups to the IESG, ultimately responsible for presenting Internet draft standards to the IAB for ratification as formal Internet standards.

Emerging discussions within this 1990 institutional structure raised concerns about a shortage of IP addresses because of rapid Internet globalization. For example, at the August 1990 IETF Vancouver meeting, participants Phill Gross, Sue Hares, and Frank Solensky projected that the current address assignment rate would deplete much of the Internet address space by March 1994.[11]

Projected address scarcity was not the only concern. IAB members also acknowledged the "rapidly growing concern internationally"[12] that American institutions controlled the distribution of Internet resources. Since the Internet's inception, there has been a central system for allocating Internet addresses. There were three general reasons for establishing central administration of these Internet numbers—scarcity, criticality, and the technical requirement of global uniqueness. The Internet's technical architecture is designed with the requirement that each Internet address be globally unique. A centralized institutional structure responsible for address allocation was intended to ensure that duplicate numbers were not assigned to different Internet devices at any given time. Additionally, these numbers

10. Gary Malkin, "The Tao of IETF, A Guide for New Attendees of the Internet Engineering Task Force," RFC 1718, November 1994.
11. Scott Bradner and Allison Mankin, "The Recommendation for the IP Next Generation Protocol," RFC 1752, January 1995.
12. Internet Architecture Board teleconference minutes, April 26, 1990. Accessed at http://www.iab.org/documents/iabmins/IABmins.1990-04-26.html.

are a critical Internet resource necessary for the Internet to function, so Internet designers wanted a trusted individual or institution to manage them. Finally, Internet addresses were a scarce resource in that a finite number were available. Although 4.3 billion represented an enormous number in the early Internet context, it was still a finite number which would conceivably require some conservation and control.

In the opening decades of the Internet, Jon Postel performed the role of distributing Internet addresses. Postel was a trusted and respected Internet technical designer who worked at the University of Southern California's (USC) Information Sciences Institute (ISI), funded by the US Department of Defense. As the task of handling Internet number assignment expanded with the growth of the Internet, others became involved and the responsibility was formalized into an institution called the Internet Assigned Numbers Authority (IANA), which remained at USC and with Jon Postel still playing a central role. In the 1990 context, IANA had delegated part of the address assignment process to SRI International's Network Information Center (called DDN-NIC), funded by the US Department of Defense. Chapter 5 will describe the detailed history of the Internet address space and associated institutions involved in address distribution. As the Internet began to globally expand, the Internet's governance structure, including the IAB, raised the concern that address assignment should be more internationally distributed rather than controlled by an American-centric institution funded by the US Department of Defense.

The two general assumptions were that the "IP address space is a scarce resource" and that, in the future, a more international, nonmilitary, and nonprofit institution might potentially assume responsibility for address allocations.[13]

After several months of discussions within the IAB, Cerf issued a recommendation to the Federal Networking Council (FNC), then the US government's coordinating body for agencies supporting the Internet, that the responsibility for assigning remaining addresses be delegated to international organizations, albeit with IANA still retaining centralized control:

With the rapid escalation of the number of networks in the Internet and its concurrent internationalization, it is timely to consider further delegation of assignment and registration authority on an international basis. It is also essential to take into consideration that such identifiers, particularly network identifiers of Class A and B

13. Ibid.

type, will become an increasingly scarce commodity whose allocation must be handled with thoughtful care.[14]

The IAB believed that the internationalization and growth of the Internet warranted a redistribution of remaining addresses to international registries but also recognized that this institutional tactic alone was insufficient for accommodating the globalization and rapid expansion of the Internet.

The IAB held a "soul searching" two-day meeting in January 1991 at the USC-ISI in Marina del Rey, California, to discuss future directions for the Internet.[15] The issue of Internet internationalization was prominent on the agenda. The IAB pondered whether it could "acquire a better international perspective," by supporting international protocols, increasing international membership in the IAB, and holding some meetings outside of the United States.[16]

The theme of Internet globalization traversed several topics including the controversial issue of export restrictions on encryption products and the divisive issue of "OSI." At the time, interoperability between different vendors' computer networking systems was not straightforward. In many networking environments, technologies developed by one manufacturer could not communicate with technologies produced by another manufacturer because they did not use common network protocols.

The Open Systems Interconnection (OSI) protocols advanced by the International Organization for Standardization, rather than by the IETF, were in contention to become the global interoperability standard, providing much needed interoperability among different products. OSI was an international standards effort sanctioned by numerous governments, particularly in Western Europe but also throughout the world. The US government, in 1990, mandated that its government procured products conform to OSI protocol specifications,[17] and even the US Department of Defense, an original proponent of TCP/IP, viewed the adoption of OSI protocols as

14. Vinton Cerf, "IAB Recommended Policy on Distributing Internet Identifier Assignment and IAB Recommended Policy Change to Internet 'Connected' Status," RFC 1174, August 1990.

15. David Clark et al., "Towards the Future Internet Architecture," RFC 1287, December 1991.

16. Internet Activities Board, Meeting Minutes, January 8–9 1991, Foreward [SIC]. Accessed at http://www.iab.org/documents/iabmins/IABmins.1991-01-08.html.

17. The United States Federal Information Processing Standards (FIPS) Publication 146-1 endorsed OSI compliant products in 1990. In 1995, FIPS 146-2 retracted this mandate.

somewhat of a global inevitability. Despite government endorsement of OSI, the competition between the protocols underlying the Internet (TCP/IP) and OSI remained unsettled in practice.

It was not entirely clear which family of network protocols, TCP/IP or OSI, would become the dominant "vendor-neutral" interoperability standard. OSI protocols had limited deployments relative to TCP/IP but had the backing of international governments and the US National Institute of Standards and Technology (NIST), and increasing investment by prominent network computing vendors such as Digital Equipment Corporation (DEC). TCP/IP was the working set of protocols supporting the growing public Internet, had garnered an increasing presence within private corporate networks, had the backing of the Internet's technical community, and had well-documented specifications, productive standards institutions, and working products. Within IAB deliberations the issues of OSI and internationalization existed contemporaneously with recognition of Internet address space constraints.

These issues surfaced together in the January 1991 joint meeting of the IAB and IESG to discuss future directions for the Internet's technical architecture. Twenty-three Internet engineers attended the meeting, including Vinton Cerf and Jon Postel.[18] The gathering was later described as "spirited, provocative, and at times controversial, with a lot of soul-searching over questions of relevance and future direction."[19]

MIT's Dave Clark commenced the meeting with an introductory presentation attempting to identify and illuminate six problem areas:

- The multiprotocol Internet
- Routing and addressing
- Getting big
- Dealing with divestiture
- New services (e.g., video)
- Security

18. The meeting minutes record the following attendees: IAB members Bob Braden, Vint Cerf, Lyman Chapin, David Clark, Phill Gross, Christian Huitema, Steve Kent, Tony Lauck, Barry Leiner, Dan Lynch, and Jon Postel; and IESG members Ross Callon, Noel Chiappa, David Crocker, Steve Crocker, Chuck Davin, Phillip Gross, Robert Hagens, Robert Hinden, Russell Hobby, Joyce Reynolds, and Gregory Vaudreuil; and FNC visitor Ira Richer, DARPA. Meeting minutes accessed at http://www.iab.org/documents/iabmins/IABmins.1991-01-08.html.
19. David Clark et al., "Towards the Future Internet Architecture," RFC 1287, December 1991.

The first area addressed the multiprotocol question of whether the Internet should support both TCP/IP and OSI protocols, a question Clark described as "making the problem harder for the good of mankind."[20] Clark identified a conflict between the ability to fulfill technical requirements promptly versus taking the time to incorporate OSI protocols within the Internet's architecture. The group discussed four alternative scenarios for the evolution of the Internet and the place of TCP/IP and OSI within this evolution. First, OSI and TCP/IP could both coexist indefinitely; second, TCP/IP could be replaced by OSI; third, OSI could fade and TCP/IP remain the protocol suite underlying the Internet; or finally, a next generation protocol suite could replace both TCP/IP and OSI.

Some meeting participants noted that, if the Internet standards institutions (IAB and IETF) redirected efforts toward bringing OSI to successful fruition, these institutions would be working on protocols over which it has no control. The overall consensus, as recorded in the meeting minutes, was that almost everyone backed the continued concurrent development of both protocol suites, TCP/IP and OSI, in the respective standards organizations. Clark also emphasized that any potential top-down mandates would not be as efficacious as grassroots approaches centered on working code. Other issues included the impact of the Internet's expansion and growing commercialization on routing and addressing architectures. The group decided that it was necessary to call an additional three-day "architecture retreat" reserved for members of the IAB and IESG to attempt to achieve some consensus about the Internet's technical and policy directions. The meeting was scheduled for June.

The promised June 1991 Internet architecture retreat included thirty-two Internet insiders from the IAB and the IESG, and some guests. These individuals represented universities, research institutions, corporations, and the US government.[21] Five IAB members, including Clark and

20. Internet Activities Board, Summary of Internet Architecture Discussion, January 8–9 1991, Appendix A, David Clark's presentation. Accessed at http://www.iab.org/documents/iabmins/IABmins.1991-01-08.arch.html.

21. Among the participants were Dave Clark, MIT; Hans-Werner Braun, SDSC; Noel Chiappa, consultant; Deborah Estrin, USC; Phill Gross, CNRI; Bob Hinden, BBN; Van Jacobson, LBL; Tony Lauck, DEC; Lyman Chapin, BBN; Ross Callon, DEC; Dave Crocker, DEC, Christian Huitema, INRIA; Barry Leiner, Jon Postel, ISI; Vint Cerf, CNRI; Steve Crocker, TIS; Steven Kent, BBN; Paul Mockapetris, DARPA; Robert Braden, ISI; Chuck Davin, MIT; Dave Mills, University of Delaware; Claudio Topolcic, CNRI. Source: RFC 1287, December 1991.

Cerf,[22] published the outcome of the retreat as an informational RFC in December 1991. This document, called "Towards the Future Internet Architecture," outlined a blueprint for the Internet's architectural development over a five- to ten-year period and sought discussion and comments from the Internet community. The blueprint established guidelines in five areas identified as the most pressing concerns for the ongoing evolution of the Internet:

- Routing and addressing
- Multiprotocol architectures
- Security architectures
- Traffic control and state
- Advanced applications.

A collective assumption was that the Internet faced an inevitable problem termed *address space exhaustion*, whereby "the Internet will run out of the 32-bit IP address space altogether, as the space is currently subdivided and managed."[23] Furthermore, the group identified this possibility, along with concerns about the burdens growth would place on the Internet's routing functionality, as the most urgent problem confronting the Internet. The group believed it should embark upon a long-term architectural transformation that would replace the current 32-bit global address space.[24]

At the time of the Internet architecture retreat, the prevailing Internet Protocol, IPv4, was a decade old. In 1981, the year IBM introduced its first personal computer, RFC 791 introduced the Internet Protocol standard. This 1981 IP specification, referred to at the time as both the DoD standard Internet Protocol and the Internet Protocol, drew from six prior iterations of IP but was its first formal version.[25]

Even though there was no official predecessor, the Internet Protocol was later named Internet Protocol version 4, or IPv4, because its function bifurcated from the Transmission Control Protocol (TCP), which previously had three versions. The Internet Protocol addresses two key networking functions: fragmentation and addressing. It specifies how to fragment

---

22. The other three co-authors were Lyman Chapin (BBN), Robert Braden (ISI), and Russell Hobby (UC Davis).
23. David Clark et al., "Towards the Future Internet Architecture," RFC 1287, December 1991, p. 4.
24. Ibid., p. 5.
25. Jon Postel, "Internet Protocol, DARPA Internet Program Protocol Specification Prepared for the Defense Advanced Research Projects Agency," RFC 791, September 1981.

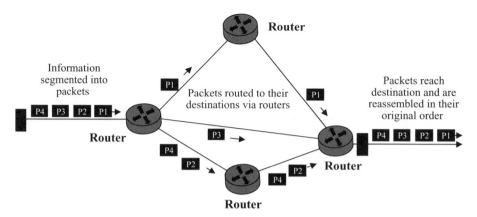

**Figure 2.1**
Packet-switching approach

and structure information into small segments, or datagrams (also called packets), for transmission over a network and reassembly at their destinations. The underlying switching approach of the Internet is called packet switching, which breaks information into packets, sequences them, and routes each packet individually over the Internet via the most available or expeditious route. Figure 2.1 illustrates the packet switching approach.

The Internet Protocol establishes how to append source and destination addresses within these packets and uses these addresses to route packets to their final destinations. Packets contain both content (or payload), such as the text of an electronic mail message, and a "header" providing control and routing information about the packet. This header information is transmitted along with the payload information. IP specifies certain fields, or spaces, within this header to describe how to fragment and then reassemble packets. The header also contains the source and destination address for the packet. Routers read a packet's destination IP address and, using routing tables, forward the packet to the next appropriate router, which, in turn, makes real-time forwarding decisions, and so forth until the packet reaches its final destination. Figure 2.2 illustrates the standard IPv4 header format.[26]

The header accompanies information sent over the Internet. As shown in figure 2.3, the first 4 bits of the header indicate the protocol version number. The next 4 bits, called IHL in the diagram, indicate the Internet

26. Ibid.

0 1 2 3 4 5 6 7 8 9 10 11 12 13 14 15 16 17 18 19 20 21 22 23 24 25 26 27 28 29 30 31

| Vers. | IHL | Type of Service | | Total Length | |
|---|---|---|---|---|---|
| Identification | | | Flags | Fragment Offset | |
| TTL | | Protocol | | Header Checksum | |
| Source Address | | | | | |
| Destination Address | | | | | |
| Options | | | | Padding | |

**Figure 2.2**
IPv4 header format

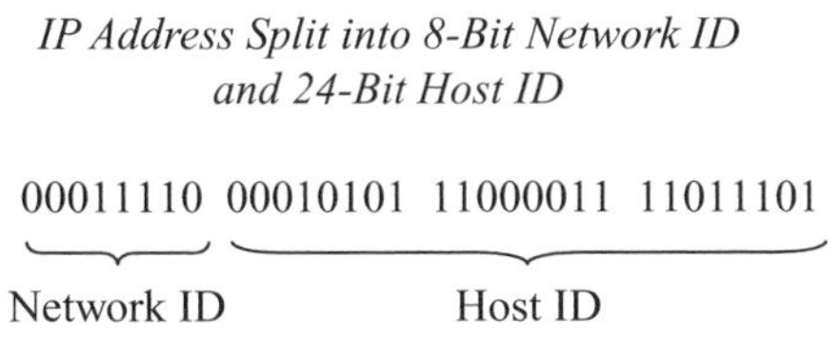

*IP Address Split into 8-Bit Network ID*
*and 24-Bit Host ID*

00011110  00010101  11000011  11011101

Network ID          Host ID

**Figure 2.3**
Network and host IDs

header length. For those interested, the technical appendix at the end of this book describes these other fields, but notice the space for "source" and "destination." These two 32-bit fields are reserved for the transmitting device's Internet address and the destination device's Internet address.

The 1981 Internet Protocol standard (formally implemented in 1983), specified an IP address as a 32-bit number, a combination of 32 0s and 1s such as the following address:

00011110000101011100001111011101

Each binary address is divided into a network prefix and a host prefix. This address division into network and host components expedites router performance. Routers store routing tables, enormous quantities of data they reference to make forwarding decisions based on the network addresses they process. Routing tables contain only network prefixes, with the exception of the end routers that directly connect to a local network.

While computing devices recognize binary sequences, the IP address format more recognizable to Internet users is in decimal format, such as 30.21.195.221. This conventional shorthand notation, called "dotted decimal format," makes 32-bit Internet addresses more numerically condensed and manageable for humans. The randomly chosen IP address listed above, 30.21.195.221, represents one out of the more than four billion theoretically available addresses. When the Internet Protocol was designed in the early 1980s, 4.3 billion represented an exorbitant number. As some within the Internet technical community would acknowledge fifteen years later, "Even the most farseeing of the developers of TCP/IP in the early 1980s did not imagine the dilemma of scale that the Internet faces today."[27]

By 1991 the technologists participating in the Internet architecture retreat agreed that the supply of more than 4.3 billion Internet addresses under the IPv4 standard would become exhausted at some future time. The retreat included a day-long breakout session for five subgroups to deliberate on the areas identified as most pressing for the Internet's architectural future. MIT's Dave Clark chaired the routing and addressing subgroup.[28] The participants identified some initial possibilities for extending the Internet address space. One alternative would retain the 32-bit address format but eliminate the requirement of global uniqueness for each address. Instead, different Internet regions would require globally unique addresses but each address could be reused in a different region. Gateways would translate addresses as information traversed the boundary between two regions. This concept was theoretically similar to frequency reuse in cellular telephony, whereby electromagnetic spectrum limitations are overcome by reusing frequencies in nonadjacent cells. When a caller moves to an adjacent cell, a hand-off process transfers the call from one frequency to another. Another alternative would expand the Internet address size, such as from 32 to 64 bits.[29] This change in address size would have exponentially increased the number of available addresses.

## Defining the Global Internet

Prior to establishing new protocol directions, the IAB believed it must first answer the question of what the Internet is. This topic arose in conjunction

27. Scott Bradner and Allison Mankin, "The Recommendation for the IP Next Generation Protocol," RFC 1752, January 1995.
28. The other members of the routing and addressing subgroup included Hans-Werner Bruan, SDSC; Noel Chiappa, Consultant; Deborah Estrin, USC; Phill Gross, CNRI; Bob Hinden, BBN; Van Jacobson, LBL; and Tony Lauck, DEC.
29. David Clark et al., "Towards the Future Internet Architecture," RFC 1287, December 1991.

with debates about whether the Internet should offer multiple protocol options, whether it should be technically homogeneous, and whether the IAB should mandate certain protocols. In the IAB's "Towards the Future Internet Architecture" document, international pressure to adopt OSI protocols as a universal computer networking standard loomed large in both the questions asked and in architectural decisions. International institutions endorsed many of the OSI protocols. The US government seemed to support OSI through its GOSIP[30] standard. The networking environments within US corporations were typically multiprotocol in 1991, with a large business usually operating some proprietary protocol networks such as IBM's Systems Network Architecture (SNA), DEC's DECnet, some TCP/IP networks, Appletalk protocols to support Apple Macintosh environments, and IPX/SPX protocols associated with Novell Netware LANs. Often these network protocol environments were isolated technical islands within large enterprises. The open question was whether TCP/IP or some other family of protocols, particularly OSI, would become the universal standard interconnecting these networks.

The technologists confronting questions about what makes the Internet the Internet were primarily based in the United States and had been in control of Internet architectural directions and responsible for Internet innovations for, in some cases, twenty years. Those involved in the Internet architecture retreat acknowledged that:

The priority for solving the problems with the current Internet architecture depends upon one's view of the future relevance of TCP/IP with respect to the OSI protocol suite. One view has been that we should just let the TCP/IP suite strangle in its success, and switch to OSI protocols. However, many of those who have worked hard and successfully on Internet protocols, products, and service are anxious to try to solve the new problems within the existing framework. Furthermore, some believe that OSI protocols will suffer from versions of many of the same problems.[31]

They presaged that both the TCP/IP and OSI protocol suites would coexist and acknowledged "powerful political and market forces" behind the introduction of the OSI suite.[32] Against the backdrop of the TCP/IP versus OSI issue, the IAB tackled the question of *what* is the Internet. The June 1991 Internet architecture retreat raised questions about whether there existed a universal criterion for what constituted the Internet

30. GOSIP: Government Open Systems Interconnection Protocol.
31. David Clark et al., "Towards the Future Internet Architecture," RFC 1287, December 1991.
32. Ibid.

or whether this definition would depend on local, particularistic environments.

First, the participants drew a sharp demarcation between the Internet as a communications system from the Internet as a community of people and institutions. Bounding the Internet with what they termed a sociological description, or "a set of people who believe themselves to be part of the Internet community" was deemed inefficacious.[33] Only its architectural constitution could define the Internet. The Internet standards community, in its attempt to define the Internet as part of its protocol selection process, believed it could, and should, devise technical definitions and assess protocol alternatives on the basis of technology with no consideration of subjective factors like culture or politics.

Within the bounds of rejecting social definitions and defining the Internet architecturally, the group found a universal description of the Internet. The group acknowledged that IP connectivity had historically defined Internet connectivity. Those using IP were on the Internet and those using another network-layer protocol were not: "This model of the Internet was simple, uniform, and—perhaps most important—testable."[34]

If someone could be PINGed (reached via IP), they were on the Internet. If they could not be PINGed, they were not on the Internet. This definition of the Internet is similar to suggestions from the philosophy of science about what constitutes a valid scientific theory, for example, scientists evaluating theories by subjecting falsifiable theories to testing and performing further evaluation by applying criteria such as uniformity and simplicity.[35] The working group evaluating alternatives to replace IPv4 also cited simplicity and universality among technical evaluation criteria. It can be argued that these criteria do not completely eliminate the subjective factors the IAB sought to exclude. For example, the definition of simplicity as a criterion is itself subjective, making an aesthetic judgment that simplistic protocol structures are preferable to complex protocols. The criterion of uniformity similarly made a subjective judgment. Many Internet stakeholders at the time, as the IAB acknowledged, wanted the choice to use either OSI network protocols or TCP/IP for Internet connectivity rather than adopt a homogeneous network protocol.

33. Ibid., p. 9.
34. David Clark et al., "Towards the Future Internet Architecture," RFC 1287, December 1991.
35. See, for example, Karl Popper, *The Logic of Scientific Discovery*, London: Routledge, 1966.

The IAB's definition also did not completely match the networking circumstances of the time. Many corporations operated large, private TCP/IP networks disjoint from the public Internet. These networks were based on IP but were isolated networks that a public Internet user could not access. Business partners and customers could, if authorized, gain access to these networks, but they were not automatically reachable via IP from the public Internet. Nevertheless, users of these large, private IP networks could PING each other, fulfilling the IAB's criteria of "being on the Internet." These private TCP/IP networks were not connected to the public Internet but would be considered part of the Internet by the IAB's definition. Additionally, some companies were technically "on the Internet" without using end-to-end IP. Some businesses in the early 1990s connected email gateways to the Internet, using protocols other than IP for internal corporate communications and only providing a gateway to the public Internet for the specific application of electronic mail. Companies accessing the public Internet through gateways would be considered not on the Internet by the IP demarcation criterion.

The IAB acknowledged the diversity of network environments and degrees of connectivity to the Internet, and grappled with a definition of the Internet tied to higher level name directories rather than IP addresses. Ultimately, though, the 1991 future Internet architecture document expressed a preference for protocol homogeneity. They considered TCP/IP "... the magnetic center of the Internet evolution, recognizing that (a) homogeneity is still the best way to deal with diversity in an internetwork, and (b) IP connectivity is still the best basis model of the Internet (whether or not the actual state of IP ubiquity can be achieved in practice in a global operational Internet)."[36]

There was also an institutional implication of this definition. With the preservation of TCP/IP, the intellectual traditions, methods, and standards' control structures within the IAB and IETF were retained. The possibility of an OSI network protocol replacing IP as the protocol tying together Internet devices had institutional control repercussions such as the International Organization for Standardization encroaching on the IAB, IETF, and IESG structures as the Internet's standards-setting and policy-making authorities. OSI was a more internationally endorsed protocol suite. For the Internet Protocol to remain the dominant protocol underpinning the Internet, it would have to meet the requirements

36. David Clark et al., "Towards the Future Internet Architecture," RFC 1287, December 1991.

of rapid international growth and, in particular, supply more Internet addresses.

The Internet's standards-setting establishment collectively embraced the objective of responding to projected international demand for more addresses but exhibited less unanimity about possible solutions. At the November 1991 Santa Fe IETF meeting held at Los Alamos National Laboratory, a new working group formed to examine the address depletion and routing table expansion issues and to make recommendations.[37]

The group, known as the ROAD group, for ROuting and ADdressing, issued specific short-term recommendations but failed to reach consensus about a long-term solution. The IESG synthesized the ROAD group's recommendations and forwarded an action plan to the IAB for consideration. Part of the IESG's recommendation was to issue a call for proposals for protocols to solve the addressing and routing problems. As the IESG chair summarized, "our biggest problem is having far too many possible solutions rather than too few."[38] Some of the options discussed in 1992 included:

- "garbage collecting,"[39] reclaiming some of the many Internet addresses that were assigned but unused;
- slowing the assignment rate of address blocks by assigning multiple Class C addresses rather than a single Class B;[40]
- aggregating numerous Class C address blocks into a larger size using a technique called classless interdomain routing (CIDR);
- segmenting the Internet into either local or large areas connected by gateways, with unique IP addresses within each area but reused in other areas; and
- enhancing or replacing IP with a new protocol that inherently would provide a larger address space.

The terms Class A and Class B used above refer to the Internet class system used at the time for distributing IP addresses in fixed blocks. The

37. The formation and objectives of the ROAD Group are described in the *Proceedings of the Twenty-Second Internet Engineering Task Force*, Los Alamos National Laboratory, Santa Fe, New Mexico, November 18–22 1991. Accessed at http://www.ietf.org/proceedings/prior29/IETF22.pdf.

38. Phillip Gross and Philip Almquist, "IESG Deliberations on Routing and Addressing," RFC 1380, November 1992.

39. From the minutes of the January 7, 1992, IAB meeting. Section 3: "Policy on Assignment and Usage of IP Network Numbers." Accessed at http://www.iab.org/documents/iabmins/IABmins.-1992-01-07.html.

40. Chapter 5 explains the Internet class system.

approximately 4.3 billion IP addresses were divided into five categories: A, B, C, D, and E. Class A, B, and C addresses were available for general distribution. Rather than requesting an ad hoc number of addresses, institutions would receive a block of addresses according to whether the assignment was designated Class A, B, or C. A Class A address assignment meant that the recipient received approximately 16 million addresses. A Class B address assignment provided roughly 65,000 addresses, and a Class C address assignment provided 256 addresses. The division into these groups had a technical rationale related to router efficiency, and Internet designers anticipated that some organizations would require large blocks of addresses while some would only need a small number.

Some of the options Internet engineers contemplated for solving addressing and routing problems never gained traction. For example, the prospect of segmenting the Internet into distinct areas separated by protocol converting gateways violated the long-standing architectural philosophy of the standards-setting community known as the "end-to-end principle."[41] Historically, Internet users trusted each other to locate important protocol functions (management, data integrity, source and destination addressing) at end nodes. Any intermediate technologies interrupting the end-to-end IP functionality would violate this principle. The possibility of reclaiming unused numbers from institutions, many of which anticipated needing them at some future date for private IP networks or public interconnection to the Internet, was also not a serious consideration, although there would later be examples of organizations voluntarily relinquishing unused address space. Plans for other options proceeded, including CIDR, more conservative assignment policies, and the development of a new protocol.

## Institutional Crisis

In the midst of questions about OSI versus TCP/IP, projected address scarcity, the growing economic importance of the Internet, and the possibility of a new protocol, the IAB was in the process of seeking greater "internationalization of the IAB and its activities."[42] The IAB had met its objective of adding some international members such as Christian Huitema of France. One of Huitema's observations was that the only IETF working

---

41. Later described by Brian Carpenter, "Architectural Principles of the Internet," RFC 1958, June 1996.
42. From the minutes of the January 7, 1992, IAB meeting. Accessed at http://www.iab.org/documents/iabmins/IABmins.1992-01-07.html.

groups with any notable non–US participation were those addressing integration with OSI applications.[43] While the IAB was seeking greater internationalization of the Internet standards process, the IETF working groups were still primarily composed of Americans. Several of these working groups were developing alternative protocol solutions to address the issues of IP address space exhaustion and routing table growth. The IESG, following the recommendations of the ROAD group, had already issued a call for proposals for new protocol solutions.

Also in 1992 a group of Internet technology veterans led by Cerf established a new Internet governance institution, the Internet Society (ISOC), a nonprofit membership-oriented institutional home and funding source for the IETF. One impetus for the establishment of this new institution was the emerging issue of liability and questions about whether IETF members might face lawsuits by those that believed Internet standards harmed them. Other drivers included a decline in US government funding of Internet standards activities and an increase in commercialization and internationalization of the Internet.

ISOC would consist of fourteen trustees with greater international representation than previous Internet oversight groups and paying corporate and individual members. At the first trustee meeting, held at an INET conference in Kobe, Japan, Lyman Chapin (the new IAB chair and also an ISOC trustee) presented a new IAB charter, "which would accomplish the major goal of bringing the activities of ISOC and the current Internet Activities Board into a common organization."[44] The trustees renamed the IAB the Internet Architecture Board (rather than Internet Activities Board), and connected the group to the newly incorporated ISOC to provide more legal protection and legitimacy. The first ISOC meeting passed a resolution assigning authority to Cerf, as ISOC president, to appoint members to a trustee nominating committee, a trustee election committee, a new committee on the Internet in developing countries, and a committee on Internet support for disaster relief.

Discussions within the Internet Society mirrored the IAB in highlighting the group's desire for greater international involvement in Internet governance, including a more formal relationship with the standards-setting body known as the International Telecommunications Union (ITU) and

43. Ibid.
44. Internet Society, Minutes of Annual General Meeting of the Board of Trustees, June 15, 1992, Kobe, Japan. Accessed at http://www.isoc.org/isoc/general/trustees/mtg01.shtml.

the establishment of Internet Society chapters throughout the world.[45] Many characteristics of this new organization distinguished ISOC from existing Internet governance institutions, including links to international standards bodies, greater international participation, direct corporate funding, and formal paying membership.

One decision the IAB made in this context created a great controversy within the Internet standards-setting community. At its June 1992 meeting in Kobe, Japan, the IAB reviewed the findings and recommendations of the ROAD group and the similar report from the IESG on the problem of Internet address space exhaustion and router table expansion. The IAB referred to the problem as "a clear and present danger" to the Internet and felt the short-term recommendations of the ROAD group, while sound, should be accompanied by the IETF endeavoring to "aggressively pursue" a new version of IP, which it dubbed "IP Version 7."[46] Rather than referring this standards development task to IETF working groups, the IAB took an uncustomary top-down step of proposing a specific protocol to replace the existing Internet Protocol, IPv4. The IAB proposed using CLNP (ConnectionLess Network Protocol), a standard that was considered part of the OSI protocol suite.

The CLNP-based proposal, called "TCP and UDP with Bigger Addresses (TUBA), A Simple Proposal for Internet Addressing and Routing,"[47] would leave higher level TCP/IP protocols (e.g., TCP and UDP) and Internet applications unchanged but would replace IP with CLNP, a protocol specifying a variable length address reaching a maximum of 20 bytes. The CLNP protocol was already a defined specification and existed, often dormant, in many vendors' products.

The IAB's decision met its objective of seeking greater internationalization of the standards process by endorsing a proposal perceived as more international. Several of the IAB members involved in the decision were directly involved in OSI protocol development and worked for companies heavily invested in OSI integration into the Internet. Ross Callon worked at DEC's Littleton, Massachusetts, facility specifically on "issues related to OSI–TCP/IP interoperation and introduction of OSI in

---

45. Ibid.

46. From the Internet Activities Board meeting minutes from the INET conference in Kobe, Japan, June 18–19, 1992. Accessed at http://www.iab.org/documents/iabmins/IABmins.1992-06-18.html.

47. See Ross Callon, "TCP and UDP with Bigger Addresses (TUBA), A Simple Proposal for Internet Addressing and Routing," RFC 1347, June 1992.

the Internet."[48] Callon had previously worked on OSI standards at Bolt Beranek and Newman (BBN). The presiding IAB chair, Lyman Chapin, worked for BBN in 1992. Chapin, also involved in standards development related to OSI, had noted the irony of formally ratifying OSI international standards but using the TCP/IP-based Internet to communicate the decision. His self-described interest was to "inject as much of the proven TCP/IP technology into OSI as possible, and to introduce OSI into an ever more pervasive and worldwide Internet."[49] IAB member Christian Huitema had also participated in OSI development, and along with Cerf believed that "with the introduction of OSI capability (in the form of CLNP) into important parts of the Internet, a path has been opened to support the use of multiple protocol suites in the Internet."[50] The IAB's CLNP-based proposal for the new Internet protocol was part of its overall internationalization objectives of integrating internationally preferred protocols into the Internet environment.

Huitema, later recollecting the IAB's CLNP recommendation, explained that he had composed the draft specification on the plane home from the Kobe meeting and that the draft went through eight revisions within the IAB over the following two weeks. Huitema recalled, "We thought that our wording was very careful, and we were prepared to discuss it and try to convince the Internet community. Then, everything accelerated. Some journalists got the news, an announcement was hastily written, and many members of the community felt betrayed. They perceived that we were selling the Internet to the ISO, that headquarters was simply giving the field to an enemy that they had fought for many years and eventually vanquished."[51]

Rank and file participants in the IETF working groups expressed outrage over the IAB's suggestion to replace IP with a new protocol based on ISO's CLNP protocol. This dismay surfaced immediately on the Internet mailing lists and at the IETF meeting held the following month. Taking into consideration that the IETF mailing lists generally contain strong opinions, the reaction to the IAB recommendations was unusually acrimonious and

48. According to RFC 1336, "Who's Who in the Internet, Biographies of IAB, IESG, and IRSG Members," published in May 1992.

49. Ibid.

50. Ibid.

51. See Christian Huitema, *IPv6 The New Internet Protocol*, Englewood Cliffs, NJ: Prentice Hall, 1996, p. 2.

collectively one of "shocked disbelief"[52] and concern that the recommen-dation "fails on both technical and political grounds."[53] The following abridged excerpts from the publicly available IETF mailing list archives (July 2–7, 1992) reflect the IETF participants' diverse but equally emphatic responses to the IAB recommendation:

I view this idea of adopting CLNP as IPv7 as a disastrous idea.
adopting CLNP means buying into the ISO standards process.
as such, we have to face the painful reality that any future changes that the Internet community wishes to see in the network layer will require ISO approval too.

Do you want to see the political equation? IPv7 = DECNET Phase 5

In voluntary systems such as ours, there is a fundamental concept of "the right-to-rule" which is better known as "the consent of the governed." Certainly the original IAB membership had a bona fide right-to-rule when it was composed of senior researchers who designed and implemented a lot of the stuff that was used. Over time, however, the IAB has degenerated under vendor and standardization influ-ences. Now, under ISO(silent)C auspices, the IAB gets to hob-nob around the globe, drinking to the health of Political Correctness, of International networking and poo-poo'ing its US-centric roots. I'm sorry, but I'm just not buying this. The Internet community is far too important to my professional and personal life for me to allow it to be sacrificed in the name of progress. For decisions this big, I'm shocked to see that IAB made the move without holding an open hearing period for opinions from the Internet community.

Procedurally, I am dismayed at the undemocratic and closed nature of the decision making process, and of the haste with which such as major decision was made.

When the IAB tells them that the IAB knows what's best—better than the best minds in this arena know, they are on very dangerous ground.

A proposed change with such extensive impact on the operational aspect of the Internet should have the benefit of considerable open discussion.

The IAB needs to explain why it believes we can adopt CLNP format and still have change control.

IETF participants considered the IAB's proposal controversial for several reasons. One of the most contentious areas concerned standards-setting procedures. The IAB's protocol recommendation had circumvented traditions within the standards-setting community in which technical

---

52. Jon Crowcroft (J.Crowcroft@cs.ucl.ac.uk) posting on the IETF mailing list, July 2, 1992.
53. Marshall Rose (mrose@dbc.mtview.ca.us) posting on the IETF mailing list, July 7, 1992.

standards percolated up from the working groups to the IESG to the IAB, not the inverse. Internet standards originated in IETF working groups and, after a period of collaboration, changes, and vetting, submitted the standard up through the institutional process for ultimate approval by the IAB. A standards decision originating in the IAB was antithetical to these traditions. Recommendations usually involved a period of public (the IETF public) review and comment prior to their selection.

Other IETF participants suggested that the IAB no longer had the legitimacy of being comprised of elders and veterans from the ARPANET days, and that new IAB members were often not involved in direct coding or standards development. They were suspicious of the recently adopted hierarchical structure that subverted the IAB under a newly formed, private, international entity—the Internet Society. Another concern was that vendors, especially DEC with its heavy investment in OSI, had undue influence in standards selection. Additionally, the new ISOC institutional structure was a departure from previous norms in that networking vendors contributed funding to the new organization.

The greatest concerns related directly to the competition between the IETF and ISO as standards bodies and to issues of power and control over standards development and change control. Some IETF participants believed that adopting an OSI standard would mean relinquishing administrative and technical control of protocols to ISO. Some questioned whether the IETF would still have "change control" and feared that protocol development would subsequently be subjected to ISO's lengthy, top-down, and complex standards-development procedures.

From a technical and procedural standpoint, some questioned why there was no comparison to the other IPv4 alternatives that IETF working groups were already developing. The IESG recommended that the community examine other alternatives for the new Internet protocol rather than uniformly pursuing the proposal based on the CLNP protocol. The backlash over the IAB's recommendation was multifaceted, involving concerns about CLNP's association with ISO, questions about whether CLNP was the best alternative, concern about the influence of corporations with a vested interest in the outcome, and alarm about the IAB's top-down procedural maneuver.

These concerns pervaded deliberations at the twenty-fourth IETF meeting convening the following month at the Cambridge, Massachusetts, Hyatt Regency adjacent to the MIT campus.[54] Participating in the more than

54. According to the *Proceedings of the Twenty-Fourth Internet Engineering Task Force*, MIT, Cambridge, MA, July 13–17 1992, compiled and edited by Megan Davres,

eighty technical working groups held during the IETF meeting were 687 attendees, a 28 percent increase over the IETF's previous meeting in San Diego. Technical and procedural challenges associated with Internet growth were the predominant topics of discussion and the meeting included a plenary session delivered by MIT's David Clark. The IETF community respected Clark as a long-time contributor to the Internet's technical architecture who had served as the ICCB's chair beginning in its 1979 inaugural year and who had also previously served as the chair of the IAB.

Clark's plenary presentation, "A Cloudy Crystal Ball, Visions of the Future," reflected the angst IETF working group participants felt about the IAB's CLNP recommendation, and ultimately articulated the philosophy that would become the IETF's de facto motto. Clark's presentation, to which he assigned the alternative title, "Apocalypse Now," attempted to examine four "forces" shaping the activities of the Internet standards-setting community: (1) new Internet services such as real-time video; (2) emerging commercial network services such as ATM (Asynchronous Transfer Mode), SMDS (Switched Multimegabit Data Service), and B-ISDN (Broadband Integrated Services Digital Network); (3) cyber-terrorists; and (4) "Us: We have met the enemy and he is. . . ." Clark's last topic, "Us," reflected upon the status and practices of the standards community. Clark compared the IAB's current role as "sort of like the House of Lords," advising and consenting to the IESG's proposals, which themselves should percolate up from the IETF working groups. Clark suggested that more checks and balances would be advantageous.

An enduring legacy of Clark's plenary presentation was his articulation of the IETF's core philosophy:

We reject: kings, presidents, and voting.
We believe in: rough consensus and running code.[55]

The phrase "rough consensus and running code" would become the IETF's operating credo. The standards community, according to Clark, had traditionally succeeded by adopting working, tested code rather than proposing top-down standards and making them work. The message was clear. Reject

---

Cynthia Clark, and Debra Legare. Accessed at http://www.ietf.org/proceedings/prior29/IETF24.pdf.

55. From David Clark's plenary presentation, "A Cloudy Crystal Ball, Visions of the Future," at the 24th meeting of the Internet Engineering Task Force, Cambridge, MA, July 1992. *Proceedings of the 24th Internet Engineering Task Force*, p. 539. Accessed at http://www.ietf.org/proceedings/prior29/IETF24.pdf.

the IAB's top-down mandate for a new protocol. The IETF's resistance to the IAB's OSI-related proposal was also evidenced by the conference's presentations and discussions of two competing protocol alternatives, PIP, the "P" Internet protocol by Bellcore's Paul Tsuchiya, and Bob Hinden's and Dave Crocker's IPAE, IP Address Encapsulation.[56]

The IAB formally withdrew its draft proposal at the IETF conference, which concluded with several outcomes: (1) the IETF would continue pursuing alternative proposals for the next generation Internet protocol rather than exclusively pursuing TUBA, (2) the Internet's core philosophy of working code and rough consensus would remain intact, (3) the standards decision process and institutional roles would be examined and revamped, and (4) as the rank and file IETF participants had desired, the influence of the more closed and more internationally oriented IAB, the influence of (some) vendors in the standards process, and the government and vendor influenced momentum of OSI protocols would be counterbalanced by grassroots solutions.

One of the institutional outcomes of the Kobe affair, at subsequent discussion on the IETF boards and at the Cambridge meeting, was a consensus decision to determine and instill a procedure for selecting members of the IESG and IAB. Immediately following the IETF meeting, Cerf, still Internet Society president and responsible for the selection of many IAB and IESG members, called for a new working group to examine issues of Internet leader selection, as well as standards processes.[57] Steve Crocker headed the working group, designated the POISED group, for Process for Organization of Internet Standards working group. At that time, Steve Crocker was a vice president at the Internet security firm Trusted Information Systems (TIS) and the IETF's area director for security. Crocker was a long-time insider in the Internet standards community and had formerly worked at USC's Information Sciences Institute and served as a research and development program manager at DARPA.

The specific charter of the new working group was to scrutinize Internet standards procedures, IAB responsibilities, and the relationship between the IAB and the IETF/IESG. For example, what should the procedures be

56. See "A PIP Presentation—The 'P' Internet Protocol" by Paul Tsuchiya of Bellcore and "IP Address Encapsulation (IPAE)" by Robert Hinden and Dave Crocker in the *Proceedings of the 24th meeting of the Internet Engineering Task Force*, p. 517. Cambridge, MA, July 1992. Accessed at http://www.ietf.org/proceedings/prior29/IETF24.pdf.
57. Steve Crocker, "The Process for Organization of Internet Standards Working Group," RFC 1640, June 1994.

for appointing individuals to the IAB? How should the standards community resolve disputes among the IETF, IAB, and IESG? Some of the working
group's conclusions[58] included term limits for IAB and IESG members and
a selection process by committees and with community input. An IETF
nomination committee would consist of seven members chosen randomly
from a group of IETF volunteers and one nonvoting chair selected by the
Internet Society.[59] The enunciation of the institutional power relations
within the Internet standards community reflexively passed the "working
code" philosophy in that the IETF attempted to retain the traditional IETF
bottom-up and participatory process it believed had worked well.

Borrowing a metaphor from the broader 1990s political discourse, Frank
Kastenholz summarized on the IETF mailing list: "the New World Order
was brought in when the IAB apparently disregarded our rules and common
practices and declared that CLNP should be IP6. They were fried for
doing that."[60] In short, the IAB recommendation and subsequent fracas
resulted in a revamping of power relations within the standards-setting
community, an articulation of its institutional values, and a demonstration
of IETF resistance to adopting any OSI protocols within the Internet's
architecture.

## Beyond Markets

After the contentious July 1992 IETF meeting, discussions about a
new protocol, referred to as Internet Protocol next generation (IPng),
dominated the IETF mailing lists and the following IETF meeting held in
Washington, DC, in November 1992. The Monday opening session commenced with competing presentations on the four proposals, at that time,
candidates to become the new Internet protocol:

- TUBA (TCP and UDP with Bigger Addresses)
- PIP ("P" Internet Protocol)

58. See the following RFCs: Internet Architecture Board and Internet Engineering
Steering Group, "The Internet Standards Process—Revision 2," RFC 1602, March
1994; Christian Huitema, "Charter of the Internet Architecture Board," RFC 1601,
March 1994; Erik Huizer and Dave Crocker, "IETF Working Group Guidelines and
Procedures," RFC 1603, March 1994; and Steve Crocker, "The Process for Organiza-
tion of Internet Standards Working Group (POISED)," RFC 1640, June 1994.

59. The process is described in Christian Huitema, "Charter of the Internet Architecture Board," RFC 1601, March 1994.

60. Frank Kastenholz posting on the IETF.ietf mailing list, March 24, 1995.

- SIP (Simple Internet Protocol)
- IPAE (IP Address Encapsulation)

TUBA, the subject of the Kobe controversy, remained on the table. This protocol, built upon the OSI-based CLNP, would replace the current Internet Protocol, IPv4, and would provide a 20-byte (160-bit) address exponentially increasing the number of devices the Internet could support. Bellcore's Paul Tsuchiya presented an alternative proposal, PIP, which would be a completely new protocol developed within the Internet's standards-setting establishment. PIP would offer a novel approach of specifying IP addresses with an unlimited address length based on dynamic requirements.

Steve Deering of Xerox PARC delivered the presentation on SIP, which he called IP Version 6. SIP would take an incremental approach of retaining the characteristics of the existing Internet Protocol but extending the address size from 32 to 64 bits. Sun Microsystem's Bob Hinden offered a technical presentation of IPAE that was actually a transition mechanism from IPv4 to a new Internet protocol, which the IPAE working group assumed would be SIP. Part of Hinden's presentation discussed how this proposed protocol differed from TUBA. A selling point of IPAE/SIP was that it would retain existing semantics, formats, terminology, documentation, and procedures and would have "no issues of protocol ownership." The competing Internet proposals, especially SIP and TUBA, were not radically different from a technical standpoint, but the question of who would be developmentally responsible for the Internet's core protocols, the established participants within the Internet's traditional standards-setting format or ISO, continued to be a distinguishing factor and an institutional concern.

At the following IETF gathering (July 1993) in Amsterdam, the first ever held outside of North America,[61] a birds of a feather (BOF) group called the IPng Decision BOF formed. A BOF group has no charter, convenes once or twice, and often serves as a preliminary gauge of interest in forming a new IETF working group.[62] The Amsterdam IPng Decision BOF, also called IPDecide, sought to discuss the decision process for the IPng selection. Two

61. Forty-six percent of the 500 attendees represented countries other than the United States, whereas previously held meetings averaged between 88 and 92 percent American attendees, according to the *Proceedings of the Twenty-Seventh Internet Engineering Task Force*, p. 10. Amsterdam, The Netherlands, July 12–16, 1993. Accessed at http://www.ietf.org/proceedings/prior29/IETF27.pdf.
62. Defined in George Malkin's "The Tao of the IETF—A Guide for New Attendees of the Internet Engineering Task Force," RFC 1391, January 1993.

hundred people attended the IPDecide BOF and consensus opinion suggested that the IETF needed to take decisive action to select IPng and that any option of letting the market decide was unacceptable. The early 1980s development of the Internet Protocol had occurred outside of market mechanisms so the idea of non–market-developed standards was not an aberrant proposition. The IPDecide BOF suggested that the marketplace already had an overabundance of protocol choices, that some architectural issues (e.g., the domain name system) could not contend with multiprotocol environments and required a single protocol, and that "the decision was too complicated for a rational market-led solution."[63]

CERN's Brian Carpenter doubted that the general market had any idea that solutions to the problem were being discussed or even that a problem existed. He believed it would take several years for the market to understand the problem and agreed with those who suggested "we still need computer science PhDs to run our networks for a while longer."[64]

The IESG created a new ad hoc working group to select IPng. The new working group tapped two Internet veterans as co–area directors (ADs) Allison Mankin of the Naval Research Laboratory, an IESG member and area director of the Internet Transport Services working group, and Scott Bradner of Harvard University's Office of Information Technology, an IESG member and area director of the Internet Operational Requirements working group.

In December 1993 Mankin and Bradner authored a formal requirements solicitation for IPng entitled RFC 1550, "IP: Next Generation (IPng) White Paper Solicitation."[65] The solicitation invited interested parties to recommend requirements IPng should meet and to suggest evaluation criteria that should determine the ultimate selection of IPng. The White Paper Solicitation promised that the submitted documents would become publicly available as informational RFCs and that the IPng working group would use this input as resource materials during the selection process.

This call for public participation and requirements input into the new Internet protocol was, in some ways, the horse behind the cart. Requirements criteria, calls for proposals, working groups, proposals, and even

63. From the minutes of the IPng Decision Process BOF (IPDecide) reported by Brian Carpenter (CERN) and Tim Dixon (RARE) with additional text from Phill Gross (ANS), July 1993. Accessed at http://mirror.switch.ch/ftp/doc/ietf/93jul/ipdecide-minutes-93jul.txt.
64. Brian Carpenter, submission to big-Internet mailing list, April 14, 1993.
65. Scott Bradner and Allison Mankin, "IP: Next Generation (IPng) White Paper Solicitation," RFC 1550, December 1993.

some evaluative comparisons of proposals had all already occurred. For example, several sets of requirements for the new protocol were circulating through the standards community. Working groups had crafted competing protocol alternatives. A formal call for proposals had been made at the contentious July 1992 IETF meeting in Cambridge.

An informational RFC Tim Dixon published in May 1993 offered one comparison of available IPng proposals. Dixon was the Secretariat of Reseaux Associés pour la Recherche Européenne (RARE), the European Association of Research Networks, which published a series of documents called RARE technical reports sometimes republished as informational RFCs. RFC 1454, "Comparison of Proposals for Next Version of IP," was a republished RARE technical document. The report compared PIP, TUBA, and SIP and concluded that the three proposals had minimal technical differences and that the protocols were too similar to evaluate on technical merit. The IPDecide BOF also had raised this issue at the Amsterdam IETF meeting, with some members suggesting that the proposals lacked significant enough technical distinctions to successfully differentiate and, even if there were differences, technical evaluation criteria were too general to argue for any one proposal.[66]

Some individuals within the IETF community were displeased with the IPng selection process. Noel Chiappa, former IETF Internet area co-director, member of the TCP/IP working group and its successor group since 1977, and formerly at MIT as a student and research staff member,[67] expressed concerns about this process. Chiappa believed a more effective approach would have been to define requirements first, or "what a new internetwork layer ought to do" and then determine how to meet those requirements.[68] Chiappa, as an independent inventor, was one of the IETF members not affiliated with a technology vendor and its products, but he had proposed his own alternative project, "Nimrod," which was not advanced as one of the IPng alternatives. Nevertheless, his criticisms illuminated several characteristics of the selection process, including the ex post facto requirements definition approach, the ongoing

66. From the minutes of the IPng Decision Process BOF (IPDecide) reported by Brian Carpenter (CERN) and Tim Dixon (RARE) with additional text from Phill Gross (ANS), July 1993. Accessed at http://mirror.switch.ch/ftp/doc/ietf/93jul/ipdecide-minutes-93jul.txt.

67. From RFC 1336, "Who's Who in the Internet: Biographies of IAB, IESG, and IRSG Members," May 1992.

68. Excerpts from Noel Chiappa posting on the info.big-internet newsgroup, May 14, 1994, Subject "Thoughts on the IPng situation. . . . "

conflict between ISO and the IETF, and the tension between grassroots versus top-down standards procedures. In short, Chiappa wrote: "That a standards body with responsibility for a key piece of the world's infrastructure is behaving like this is frightful and infuriating."[69]

Instead of technically differentiating the proposals, the RARE report suggested a political rational for a formal selection process: "the result of the selection process is not of particular significance, but the process itself is perhaps necessary to repair the social and technical cohesion of the Internet Engineering Process."[70]

Dixon highlighted the ongoing tension about OSI permeating the IPng selection, suggesting that TUBA faced a "spurious 'Not Invented Here' prejudice,"[71] on one hand, and warning that the new protocol ironically faced the danger of what many perceived as the shortcomings of the OSI standards process: "slow progress, factional infighting over trivia, convergence on the lowest common denominator solution, lack of consideration for the end-users."[72] The IETF BOF group raised another rationale for conducting a formal protocol evaluation process, citing the possibility of "potential legal difficulties if the IETF appeared to be eliminating proposals on arbitrary grounds."[73] Within the context of what some considered technically similar proposals, ongoing anxiety about OSI, fear of possible legal repercussions, and rapid global Internet growth, the IETF issued its White Paper Solicitation for requirements the next generation Internet protocol should meet. Mankin's and Bradner's brief, six-page solicitation invited interested parties to submit documents detailing requirements for IPng that could be used by the IPng area working groups to complete the selection process for the new protocol. Some questions in the solicitation included: what was the required time frame for IPng; what security features should the protocol include; what configuration and operational parameters are necessary; and what media, mobility, topology, and marketplace requirements should IPng meet?

69. Ibid.

70. Tim Dixon, "Comparison of Proposals for Next Version of IP," RFC 1454, May 1993.

71. Ibid.

72. Ibid.

73. From the minutes of the IPng Decision Process BOF (IPDecide) reported by Brian Carpenter (CERN) and Tim Dixon (RARE) with additional text from Phill Gross (ANS), July 1993. Accessed at http://mirror.switch.ch/ftp/doc/ietf/93jul/ipdecide-minutes-93jul.txt.

Bradner and Mankin received twenty-one responses to their White Paper Solicitation. Three of these submissions came from companies in industries, at the time, considered poised to become future "information superhighway" providers: the cable television industry, the cellular telephone industry, and the electric power industry.[74] These companies and industries, as potentially new Internet providers, obviously had a vested interest in the standard with which their services would likely comply. Other submissions addressed specific military requirements, corporate user requirements, and security considerations. Several submissions were recapitulations of the actual protocol proposals currently competing for IPng status.

## US Corporate Customer Perspective

One area of IPng accord within the Internet standards-setting community continued to be the espousal of the following philosophy: "the IETF should take active steps toward a technical decision, rather than waiting for the 'marketplace' to decide."[75]

Nevertheless, some of the White Paper responses reflected market requirements of large corporate Internet users, which comprised a major market sector of an increasingly commercialized Internet industry. Large corporate Internet users did not uniformly believe in the need for a next generation Internet protocol. Historian of technology Thomas Hughes suggests new technology advocates err severely in underestimating the inertia and tenacity of existing technological systems.[76] Once developed and installed, technological systems acquire conservative momentum. This momentum arises from such characteristics as financial investments, political and institutional commitments, personal stake, knowledge base, and installed material conditions. Hughes's examples of conservative momentum primarily address large systems developers, describing how technological systems reflect powerful interests with substantially vested capital and human resources that a significant system change might jeopardize.[77] In the case

74. See: Ron Skelton, "Electric Power Research Institute Comments on IPng," RFC 1673, August 1994; Mark Taylor, "A Cellular Industry View of IPng," RFC 1674, August 1994; and Mario Vecchi, "IPng Requirements: A Cable Television Industry Viewpoint," RFC 1686, August 1994.

75. Bullet point presented by the IETF chair in a meeting entitled "IPDecide BOF" at the 1993 IETF, Amsterdam.

76. Thomas Hughes, *American Genesis: A History of the American Genius for Invention*, New York: Penguin Books, 1989, p. 459.

77. Ibid., p. 460.

of a new Internet protocol, US corporate users represented a conservative foundation for IPv4. US corporate Internet users generally had ample IP addresses and substantial investment in IPv4 capital and human resources.

Boeing Corporation's response to the White Paper Solicitation sought to summarize the US corporate user view: "Large corporate users generally view IPng with disfavor."[78] Boeing suggested that Fortune 100 corporations, then heavy users of private TCP/IP networks, viewed the possibility of a new protocol, IPng, as "a threat rather than an opportunity."[79] In the early 1990s large US corporations primarily operated mixed network protocol environments rather than a single network protocol connecting all applications and systems. Corporations wanted a single, interoperable suite of protocols, but it was not yet clear which of several alternatives, if any, would meet this requirement. The Boeing Corporation's White Paper response acknowledged that it used at least sixteen distinct sets of protocols within its corporate networks. Typifying large corporate network users in this era, Boeing had an installed base of older network protocol suites like SNA, DECnet, AppleTalk, IPX/SPX, and also private TCP/IP networks. Many TCP/IP implementations within large business environments supported internal networks and did not connect to the Internet. Each protocol environment required distinct technical skills, equipment, and support infrastructures.

The prevailing trend was as a result to reduce the number of network protocol environments rather than expand them, or as the Boeing response summarized, it came as "a basic abhorrence to the possibility of introducing 'Yet Another Protocol' (YAP)."[80] TCP/IP implementations relied entirely on the prevailing IPv4 protocol, and Boeing suggested its TCP/IP network was approaching the point of interconnecting 100,000 host computers. Even if the global Internet universally adopted a new Internet protocol, Boeing believed it could deploy an application level gateway at the demarcation point between its network and the Internet to convert between IPv4 and the new IPng. The one possible economic rationale for adopting a new protocol would be market introduction of "killer apps" relying solely on IPng. The introduction of greater TCP/IP security might also help justify laboriously converting 100,000 computing devices to a new protocol.

78. Eric Fleischman, "A Large Corporate User's View of IPng," RFC 1687, August 1994.
79. Ibid.
80. Ibid.

Boeing also acknowledged prevailing tension between OSI and TCP/IP and suggested that any ability of IPng to foster a convergence between the two protocol suites would make IPng more desirable. It sold products in a global marketplace, often to government customers. Support of a protocol integrated with OSI could prove advantageous in competitive bids for contracts from governments supporting OSI. Additionally, an OSI-based protocol was beginning to replace proprietary network protocols for air-to-ground and ground-to-ground communications so that any OSI convergence IPng could achieve would make the protocol more economically appealing. Boeing further suggested that any IPng approach should provide an eventual integration between what it termed Internet standards versus international standards. Even if IPng could achieve an integration with OSI, offer new applications, or add functionality such as improved security, Boeing and other corporate users wanted IPng to coexist with the massive installed base of IPv4 for the foreseeable future.

The one potential rationale for deploying a new protocol not cited by Boeing was the need for more IP addresses. In other words, "address depletion doesn't resonate with users."[81] According to Internet address distribution records, at the time, Boeing controlled 1.3 million unique addresses.[82] Large American corporate Internet users generally had abundant Internet address reserves, and as Boeing suggested, only a new "killer app" requiring IPng would motivate them to replace their current implementations with a new Internet protocol.

IBM's White Paper response reinforced the extent of conservative momentum behind the IPv4 standard, suggesting "IPv4 users won't upgrade to IPng without a compelling reason."[83] Similarly, BBN, the developer of ARPANET's original Interface Message Processors, noted that the IPng effort was "pushing" network technology. The BBN response stressed that marketplace demands should drive the development of IPng and questioned whether IPv4 users would ever have a compelling justification to upgrade to a new protocol.[84]

81. Ibid.
82. Boeing held at least twenty distinct Class B address blocks and eighty Class C address blocks. Each Class B address block contains more than 65,000 addresses, and each Class C contains 256 addresses. So Boeing controlled at least 1.3 million IP addresses. Source for address assignment records: Sue Romano, Mary Stahl, Mimi Recker, "Internet Numbers," RFC 1117, August 1989.
83. Edward Britton and John Tavs, "IPng Requirements of Large Corporate Networks," RFC 1678, August 1994.
84. John Curran, "Market Viability as a IPng Criteria," RFC 1669, August 1994.

In contrast, companies without significant investment in IPv4 or positioned to profit from the availability of more addresses or the development of new products and services embraced the idea of a new protocol. This was especially true among industries that were potential new entrants into the Internet service provider market. The early 1990s growth and commercialization of the Internet, as well as discussions of a multimedia "global information superhighway" or "National Information Infrastructure" within the Clinton administration and in the media, drew attention to the economic potential for non–Internet network service providers to enter the increasingly lucrative Internet services marketplace.

The new Internet application, the World Wide Web, spurred significant Internet growth in the early 1990s. US-based corporations embraced the capabilities of this new application to instantly reach customers and business partners. The Clinton administration established an Internet presence with its own website and electronic mail addresses for the President, Vice President Al Gore, and First Lady Hillary Clinton. In September 1993 Gore and Secretary of Commerce Ron Brown formally announced a National Information Infrastructure (NII) initiative, an expansive economic and social project to promote a national network linking together a variety of network infrastructures and, by 2000, at a minimum "all the classrooms, libraries, hospitals, and clinics in the United States."[85] Also called the information superhighway, the NII initiative did not directly refer to the Internet but to a more broad amalgamation of telecommunications networks, entertainment, and cable systems. The initiative both highlighted possibilities for Internet expansion and intimated that alternative infrastructures, especially cable systems, might provide separate services competing with or complementing the Internet.

In 1993 there was little convergence of different information types over a common medium. Telephone networks and cellular systems supported voice, computer networks supported data, and cable companies transmitted video. The promise of integrating these services over a converged, multimedia service represented an enormous opportunity, and several of the White Paper responses reflected this interest. Companies in industries not supporting data transmission, and that had never been closely involved in Internet standards development, were interested in a new protocol, IPng, as a way to suddenly compete with existing Internet and data providers like major national telephone companies and new ISPs.

85. National Information Infrastructure White Paper, "Administration White Paper on Communications Act Reforms." Accessed at http://ibiblio.org/pub/academic/political-science/internet-related/NII-white-paper.

Cable companies envisioned opportunities to become providers of converged services, and one much touted promise of the "information superhighway" was video-on-demand, the ability to order a movie in real time over a network through a set-top box connected to a television or computer. The emergence of this service outside of cable systems, such as through an ISP, would threaten the cable industry. This interest to expand into the data services market, or at least protect its core market, was reflected in Time Warner Cable's response to the IPng White Paper Solicitation, "IPng Requirements: A Cable Television Industry Viewpoint."[86] The response described the potential for cable television networks, because of their ubiquity and broadband capacity, to become the dominant platform for delivery of interactive digital services supporting integrated voice, video, and data information.

At the time only a small percentage of American consumers had home Internet access, and there was no interactive network combining video and data transmissions. Time Warner was building a highly publicized, experimental broadband network in Orlando, Florida, promising to integrate video, voice, and data services. This offering would involve a network based on Asynchronous Transfer Mode (ATM) networks connected to a "set-top" box linked to the consumer's television. The purpose of the Time Warner Cable White Paper response was to position itself, and the cable industry generally, as dominant future providers of converged "information superhighway" services and to embrace IPng as a potential protocol supporting broadband interactive cable service. IP, as a network protocol for addressing and routing, actually would have no relationship or ability to directly facilitate the convergence of voice, video, and data but was nevertheless embraced as a way to provide more addresses, therefore reaching more consumers. IPng effectively presented a late entrant opportunity to enter the Internet marketplace and become involved in the Internet standards process.

The cellular industry was another sector not yet involved in Internet services but hoping to become competitive through the potential of converged voice and data services. Mark Taylor of McCaw Cellular Communications, Inc., responded on behalf of the Cellular Digital Packet Data (CDPD) consortium of cellular providers. The primary requirements of the digital cellular consortium were mobility, the ability to "operate anywhere anytime," and scalability, meaning "IPng should support at least tens or hundreds of billions of addresses."[87]

86. Mario Vecchi, "IPng Requirements: A Cable Television Industry Viewpoint," RFC 1686, August 1994.
87. Mark Taylor, "A Cellular Industry View of IPng," RFC 1674, August 1994.

The Electric Power Research Institute (EPRI) also submitted an interesting response to the IPng White Paper Solicitation on behalf of the electric power industry. The EPRI, a nonprofit research and development institution representing seven hundred utility companies, specifically linked the future of IP to the National Information Infrastructure and compared its importance to standards for railroads, highways, and electric utilities. The EPRI response suggested that, while the electric power industry currently used TCP/IP protocols, it was pursuing a long-term strategy of employing OSI protocols. In short, the requirements of the electric power industry "are met more effectively by the current suite of OSI protocols and international standards under development."[88] One of the reasons EPRI stated that it preferred OSI standards was that it believed the NII should have an international perspective. Another reason for endorsing OSI protocols was that the EPRI had already, according to its White Paper submission, developed and invested in industry-specific communications standards and services based on OSI.

### ISO Standard and IETF Standard Compared

Upon completion of the White Paper Solicitation process, who would decide which protocol proposal would become IPng? Bradner and Mankin, as the IPng area directors, would make the final recommendation to the IESG for approval. Additionally, the IESG also established an "IPng Directorate" to function as a review body for the proposed alternatives that existed prior to the White Paper Solicitation process. The IPng Directorate, over the course of the selection process, included the following individuals:[89] J. Allard, Microsoft; Steve Bellovin, AT&T; Jim Bound, Digital; Ross Callon, Wellfleet; Brian Carpenter, CERN; Dave Clark, MIT; John Curran, NEARNET; Steve Deering, Xerox PARC; Dino Farinacci, Cisco; Paul Francis, NTT; Eric Fleischmann, Boeing; Robert Hinden, Sun Microsystems; Mark Knopper, Ameritech; Greg Minshall, Novell; Yakov Rekhter, IBM; Rob Ullmann, Lotus; and Lixia Zhang, Xerox.

Bradner and Mankin later indicated these individuals were selected for diversity of technical knowledge and equitable representation of those involved in each IPng proposal working group.[90] The group represented

---

88. Ron Skelton, "Electric Power Research Institute Comments on IPng," RFC 1673, August 1994.

89. Scott Bradner and Allison Mankin, "The Recommendation for the IP Next Generation Protocol," RFC 1752, January 1995.

90. Ibid.

numerous technical areas spanning routing, security, and protocol architectures and so exhibited diversity in this sense. By other measurements the IPng Directorate could not be described as diverse. The majority (88 percent) of IPng Directorate members represented software vendors (Microsoft, Novell, Lotus, Sun Microsystems), hardware vendors (Digital, Wellfleet, Cisco, IBM) or their research arms (Xerox PARC), and service providers (AT&T, NEARNET, NTT, Ameritech). These corporations would presumably incorporate the new standard, once selected, into their products and therefore had an economic stake in the outcome. Most of the corporations represented on the IPng Directorate were based in the United States. The only academician on the IPng Directorate was MIT's David Clark, again a respected long-time member of the Internet's technical community. The majority of members were male and only one member represented Internet users, and only corporate Internet users.

There was no direct representation on the IPng Directorate of the US government or any other government. Many participants in the 1990s standards-setting community had corporate organizational affiliations so the IPng Directorate composition was not surprising. One "rule at start" for the IPng Directorate was that no IESG or IAB members would participate, although Directorate members Brian Carpenter and Lixia Zhang were both also IAB members. Bradner and Mankin emphasized that the IAB would implicitly not participate in the ultimate approval process, a ground rule emphasizing the IAB's diminished standards-setting credibility after the Kobe affair.[91]

By the final IPng evaluation process, three proposals were in contention to become the next generation Internet protocol: SIPP (Simple Internet Protocol Plus), CATNIP (Common Architecture for the Internet), and TUBA (TCP and UDP with Bigger Addresses); see table 2.1. The proposed protocols shared two major functional approaches: all would provide larger address fields allowing for substantially more addresses, and all would become a universal protocol. Although the proposals had technical differences, two distinguishing characteristics were who was behind the development of the standard and whether it would preserve IP or discard it.

Protocol ownership and control continued to remain a significant concern. Internet legal scholar Larry Lessig has said: "the architecture of cyberspace *is* power in this sense; how it is could be different. Politics is about how we decide. Politics is how that power is exercised, and by

91. Scott Bradner and Allison Mankin, "IPng Area Status Report," 29th IETF Conference, Seattle, March 28, 1994.

**Table 2.1**

Final IPng valternatives

|  | CATNIP | SIPP | TUBA |
|---|---|---|---|
| *Formal name* | Common Architecture for the Internet | Simple Internet Protocol Plus | TCP/UDP with Bigger Addresses |
| *Working Group chair/s* | Vladimir Sukonnik | Steve Deering, Paul Francis, Robert Hinden (past WG chairs: Dave Crocker, Christian Huitema) | Mark Knopper Peter Ford |
| *Protocol approach* | New network protocol integrating Internet, OSI, and Novell protocols 160-bit addresses | Evolutionary step from IPv4 | Replacement of IPv4 with ISO protocol CLNP 160-bit addresses |
| *Address format* | OSI NSAP address space | 64-bit addresses space | OSI NSAP address |

whom."[92] Janet Abbate elaborates that "technical standards are generally assumed to be socially neutral but have far-reaching economic and social consequences, altering the balance of power between competing businesses or nations and constraining the freedom of users."[93]

The SIPP proposal was a collaborative merging of previous proposals, IPAE, SIP, and PIP, and championed by experienced IETF insiders Steve Deering of Xerox PARC and Bob Hinden of Sun Microsystems. Sun Microsystems was closely associated with TCP/IP environments and so had a vested interest in maintaining IP as the dominant network level protocol. SIPP was the only proposal preserving IP and part of the technical specification called for expanding the address size from 32 bits to 64 bits. CATNIP would be a completely new protocol with the objective of providing a convergence of the Internet, ISO protocols, and Novell products. In other words, it would integrate three specific protocols: CLNP, IP, and IPX. CATNIP would actually use the OSI-based Network Service Access Point (NSAP) format for addresses. Robert Ullman of Lotus Development Corporation and Michael McGovern of Sunspot Graphics authored the CATNIP proposal and were explicit in their endorsement of ISO standards and their

92. Larry Lessig, *Code and Other Laws of Cyberspace*, New York: Basic Books, 1999, p. 59.

93. Janet Abbate, *Inventing the Internet*, Cambridge: MIT Press, 1999, p. 179.

belief that convergence with ISO protocols was an essential requirement for the new protocol.

The TUBA proposal was an even greater endorsement of ISO as a standards body because it specified the ISO-approved protocol, CLNP. TUBA would completely displace IP, would provide a 20-byte (160-bit) address, and, like CATNIP, would use the ISO-supported NSAP address space.

The IPng Directorate considered CATNIP not adequately specified and the deliberations on the Internet mailing lists indicated a binary choice between TUBA and SIPP. The decision for a new protocol was reduced, in effect, to a choice between an extension of the prevailing IETF Internet Protocol (SIPP) and an OSI protocol (TUBA).

There appeared to be some degree of inevitability that the selected protocol would be an extension of IPv4. The presumption that IP would triumph permeated several aspects of the selection's lexicon and process. First, an asymmetrical aspect of the selection process was the name of the future protocol—IPng, IP next generation. The nomenclature referring to the new protocol specification reflected the initial assumption that the new protocol would be an extension of the existing protocol, IP.

Second, the IAB's 1991 "Towards the Future Internet Architecture" document (RFC 1287) had concluded that IP was the one defining architectural component of the Internet, with those using IP considered on the Internet and those using another network-layer protocol not on the Internet. Selecting a different network-layer protocol would make the Internet not the Internet, by this definition.

Finally, the presumption that the new protocol would be an extension and modification of IP was present in the evaluation criteria for IPng, as the following chronology suggests. Bradner and Mankin stated that Craig Partridge of BBN and Frank Kastenholz of FTP Software submitted the "clear and concise set of technical requirements and decision criteria for IPng"[94] in their document "Technical Criteria for Choosing IP the Next Generation (IPng)." The authors explained that their derivation of criteria emanated from several sources, including discussions on the Internet mailing lists, IETF meetings, and from IPng working group meetings.[95] The 1995 "Recommendation for IPng," RFC 1752, contained a lengthy summary of nineteen selection criteria that Partridge and Kastenholz had defined

94. Scott Bradner and Allison Mankin, "The Recommendation for the IP Next Generation Protocol," RFC 1752, January 1995.

95. Craig Partridge and Frank Kastenholz, "Technical Criteria for Choosing IP the Next Generation (IPng)," RFC 1726, December 1994.

earlier in RFC 1726.[96] The list of criteria in the "Recommendation for IPng" document did not include the following criterion from the original list that Partridge and Kastenholz devised:

*One Protocol to Bind Them All*   One of the most important aspects of the Internet is that it provides global IP-layer connectivity. The IP layer provides the point of commonality among all nodes on the Internet. In effect, the main goal of the Internet is to provide an IP Connectivity Service to all who wish it.[97]

The requirement for global IP connectivity was the only evaluation criterion not carried over from the twenty original "Technical Criteria for Choosing IP the Next Generation" document into the explanation, in "Recommendation for IPng," for how the proposals were evaluated. This technical criterion carried a SIPP predisposition as the only proposal based on IP. The nineteen officially sanctioned technical evaluation criteria for the new protocol, omitting the requirement for global IP connectivity, included the following (paraphrased):

- Completeness   Be a complete specification.
- Simplicity   Exhibit architectural simplicity.
- Scale   Accommodate at least $10^9$ networks.
- Topological flexibility   Support a diversity of network topologies.
- Performance   Enable high-speed routing.
- Robust service   Provide robust service.
- Transition   Include a straightforward transition from IPv4.
- Media independence   Operate over a range of media using a range of speeds.
- Datagram service   Accommodate unreliable delivery of datagrams.
- Configuration ease   Enable automatic configuration of routers and Internet hosts.
- Security   Provide a secure network layer.
- Unique names   Assign globally unique identifiers to each network device.
- Access to standards   Provide freely available and distributable standards with no fees.
- Multicast support   Support both unicast and multicast transmissions.
- Extensibility   Be able to evolve to meet future Internet needs.
- Service classes   Provide service according to classes assigned to packets.
- Mobility   Support mobile hosts and networks.
- Control protocol   Include management capabilities like testing and debugging.
- Tunneling support   Allow for private IP and non–IP networks to traverse network.

The overall selection process, and even the specific technical evaluation criteria, reflected a tension between what the participants considered

96. Ibid.
97. Ibid.

evaluating the proposals technically versus evaluating proposals politically. Bradner and Mankin recognized and acknowledged the politics involved in the decision, characterizing it as pressure for convergence with ISO versus pressure to resist ISO standards and retain protocol control within the IETF. As they described in their IPng Area Status Report at the IETF meeting in Seattle on March 28, 1994, the pressure for convergence with the ISO was something the working group had to understand but must "dismiss as not a technical requirement."[98]

The selection process seemed to exhibit some asymmetry about what was considered political, with positions advocating technical convergence with OSI standards deemed political but positions against convergence (i.e., preserving IP) considered technical. The 1991 Internet architecture document had acknowledged "powerful political and market forces"[99] behind the introduction of the OSI suite, and this sentiment appeared to persist years later during the IPng selection process that considered "convergence" not a technical issue but a political issue. The process appeared to define the ISO preference for protocol convergence as a political bias and define preferences for a non–ISO protocol as technical criteria.

The political issue the IPng Directorate directly acknowledged and addressed related to control over the standard. The IETF wanted to retain protocol ownership (i.e., change control), even if they selected the ISO-based protocol, TUBA. This issue represented an area of discord even within the TUBA working group, with some arguing that only ISO should control the standard and others believing the IETF should have authority to modify the standard. This battle for change control over the new standard permeated deliberations within the working groups and the IPng Directorate, was reflected in the mailing list forums, and even in draft proposals competing groups issued. For example, the proposed CATNIP alternative included the following statement: "The argument that the IETF need not (or should not) follow existing ISO standards will not hold. The ISO is the legal standards organization for the planet. Every other industry develops and follows ISO standards. ISO convergence is both necessary and sufficient to gain international acceptance and deployment of IPng."[100]

98. From Scott Bradner and Allison Mankin, IPng Area Status Report given at IETF 29, Seattle, WA, March 28, 1994. Accessed at http://www.sobco.com/ipng/presentations/ietf.3.94/report.txt.

99. David Clark et al., "Towards the Future Internet Architecture," RFC 1287, December 1991.

100. Michael McGovern and Robert Ullman, "CATNIP: Common Architecture for the Internet," RFC 1707, October 1994.

Many expressed the opposite sentiment and concern about the possibility of relinquishing protocol control to ISO was especially prevalent on the big Internet mailing list, the forum used to discuss the proposals and the site where Mankin and Bradner posed questions to the IETF standards community. For example, one IETF participant declared that "the decisions of ISO are pretty irrelevant to the real world, which is dominated by IETF and proprietary protocols."[101]

A significant factor in the evaluation process was whether the IETF would retain control of the protocol or whether ISO would assume change control.

## Announcement of IPv6

At the opening session of the thirtieth meeting of the IETF in Toronto, Canada, Bradner and Mankin presented their recommendation that SIPP, with some modifications, become the basis for IPng. SIPP would represent an evolutionary step from the existing Internet Protocol and would preserve control of the new standard within the IETF. More than 700 people attended the IETF meeting, with the high attendance rate attributable to excitement about the protocol announcement and an increase in press representation.[102] IANA formally assigned the version number "6" to IPng so the new protocol would be named IPv6. IPv4 was the prevailing version of IP and number 5 had already been allocated to an experimental protocol. The next version number available was 6. (The nomenclature "IPv7" for the Kobe protocol had erroneously skipped over 6.)

Mankin and Bradner recounted how the IPng Directorate had identified technical flaws in each proposal; see table 2.2. The Directorate had dismissed CATNIP as an insufficiently developed protocol. The general technical assessment of TUBA and SIPP suggested that "both SIPP and TUBA would work in the Internet context"[103] despite technical weaknesses in each approach. Yet the assessment of TUBA was also "deeply divided."[104] The Directorate identified some technical weaknesses in the CLNP protocol, the centerpiece

101. Donald Eastlake, posting on big-Internet mailing list, September 14, 1993.
102. According to the Director's Message, *Proceedings of the Thirtieth IETF*, Toronto, Ontario, Canada, July 25–29, 1994. Accessed at ftp://ftp.ietf.org/ietf-online-proceedings/94jul/directorsmessage.txt.
103. Scott Bradner and Allison Mankin, "The Recommendation for the IP Next Generation Protocol," RFC 1752, January 1995.
104. From the text version of the IPng presentation Scott Bradner and Allison Mankin made at the IETF meeting in Toronto on July 25, 1994. Accessed at http://www.sobco.com/ipng/ presentations/ietf.toronto/ipng.toronto.txt.

**Table 2.2**
Protocol selection criteria

|  | Proposals evaluated against technical requirements | | |
|---|---|---|---|
|  | CATNIP | SIPP | TUBA |
| Complete specification | No | Yes | Mostly |
| Simplicity | No | No | No |
| Scale | Yes | Yes | Yes |
| Topological flex | Yes | Yes | Yes |
| Performance | Mixed | Mixed | Mixed |
| Robust service | Mixed | Mixed | Yes |
| Transition | Mixed | No | Mixed |
| Media independent | Yes | Yes | Yes |
| Datagram | Yes | Yes | Yes |
| Configuration ease | Unknown | Mixed | Mixed |
| Security | Unknown | Mixed | Mixed |
| Unique names | Mixed | Mixed | Mixed |
| Access to standards | Yes | Yes | Mixed |
| Multicast | Unknown | Yes | Mixed |
| Extensibility | Unknown | Mixed | Mixed |
| Service classes | Unknown | Yes | Yes |
| Mobility | Unknown | Mixed | Mixed |
| Control protocol | Unknown | Yes | Mixed |
| Tunneling | Unknown | Yes | Mixed |

Source: From RFC 1752, "The Recommendation for IPng."

of the TUBA proposal, but division also remained about IETF ownership of the protocol. Two of Mankin's and Bradner's comments reflected this division, "TUBA is good because of CLNP. If not CLNP, it is a new proposal" and "if TUBA becomes the IPng, then the IETF must own TUBA."

If the IETF modified CLNP, some believed this would negate the advantage of CLNP's installed base and would diminish the possibility of a successful convergence between ISO and IETF standards. If IETF could not modify CLNP, it would lose control of the Internet. Christian Huitema, an IAB member involved in the SIPP working group, later summarized his assessment of the reason TUBA was not selected: "In the end this proposal failed because its proponents tried to remain rigidly compatible with the original CLNP specification."[105]

105. Christian Huitema, *IPv6: The New Internet Protocol*, Englewood Cliffs, NJ: Prentice Hall, 1996, p. 5.

With CATNIP and TUBA eliminated, SIPP became IPng, now renamed IPv6. Members of the IPng Directorate also identified numerous technical issues with SIPP, including considerable operational problems with IPAE (the IPv4 to IPng transition mechanism), inadequate address size, and insufficient support for autoconfiguration, mobility, and security. A significant modification to SIPP was that the new SIPP-based protocol, IPv6, would have 128-bit addresses rather than 64-bit addresses.

A new IPng working group would form to develop the new IPv6 specifications and resolve open or unfinished issues. Steve Deering, the primary SIPP architect, and Ross Callon, who had been a proponent of TUBA, became co-chairs of the new working group, illustrating a conciliatory attempt to unify the TUBA and SIPP bases. The IESG approved the IPv6 recommendation, which became a "proposed standard," in accordance with the IETF's conventional nomenclature, on November 17, 1994.

The most significant difference between IPv4 and IPv6 was the expansion of the Internet address length from 32 to 128 bits, increasing the number of available addresses from approximately 4.3 billion to $3.4 \times 10^{38}$ addresses. This address length expansion represented only one technical change in the protocol. Another modification was a significant simplification of the header format. Recall that headers contain the control information preceding content transmitted over a network, analogous to the function of an envelope for mailing a letter. Header content includes information such as source address, destination address, and payload length. The IPv6 header specification eliminated some information to keep the header size as compact as possible, especially considering its larger address size. To illustrate the header simplification IPv6 provided, IPv6 addresses are four times longer than IPv4 addresses but the IPv6 header is only two times longer than the IPv4 header. Another distinction between the newly selected IPv6 protocol and IPv4 included support for autoconfiguration, an attempt to simplify the process of adding IPv6 nodes into a "plug and play" scenario whereby users could plug in a computer and have it connected via IPv6 without extensive intervention. The specification also included a format extension designed to encourage encryption use. As the IPv6 specification stated, "Support for this (security) extension will be strongly encouraged in all implementations."[106]

The 1994 decision to proceed with a SIPP-based IPv6 concluded two years of deliberations about selecting a new protocol. The selection retained IP, though modified, as the dominant network-layer protocol for the Internet

106. Scott Bradner and Allison Mankin, "The Recommendation for the IP Next Generation Protocol," RFC 1752, January 1995.

and settled the issue of who would control the next generation Internet protocol.

Bradner and Mankin closed their IETF plenary presentation recommending IPv6 with the following two quotes and a concluding sentiment:

In anything at all, perfection is finally attained not when there is no longer anything to add, but when there is no longer anything to take away.

—Antoine de Saint-Exupéry

Everything should be made as simple as possible, but not simpler.

—Albert Einstein

IETF work is trying to find the right understanding of the balance between these two goals. We think we have done that in IPng.[107]

### Themes in Protocol Development

This chapter described how the institutional trajectory leading to the IPv6 standard reflected tensions among an expanding sphere of Internet stakeholders. The issue of protocol selection was also an issue of power selection. Internet standards development can easily be viewed as "just a technical design decision," but political and economic interests also enter the process. In the context of Internet globalization, a significant area of conflict underlying the selection of IPv6 was the question of who would control the direction of the Internet's architecture.

Internet architects devised the very definition of the Internet in technical terms—those who used IP were on the Internet and those who used a different protocol were not. The Internet architects selecting the new protocol also stressed that only technical requirements would factor into the selection of the new standard. These same architects also recognized that the ability to control the Internet's technical architecture in the future depended on which protocol alternative was selected.

If a different alternative had been selected, control of the Internet's technical direction could have shifted from the IETF to ISO. The IETF was the institution traditionally responsible for the Internet's technical architecture and was made up of individuals who had been responsible historically for inventing the Internet's protocols. At the time the majority of individuals involved in the IETF were American. In contrast, ISO was a

---

107. From the text version of the IPng presentation Scott Bradner and Allison Mankin made at the IETF meeting in Toronto on July 25, 1994. Accessed at http://www.sobco.com/ipng/ presentations/ietf.toronto/ipng.toronto.txt.

more international organization and was advocating a set of technical standards (OSI protocols) that were in competition with the Internet's TCP/IP protocols to become the universal solution to interoperability among heterogeneous networks and computing devices. If an ISO-developed protocol had been selected as the new Internet protocol, the ability to enact future changes to the Internet's key protocol would likely have rested with ISO, not the IETF. The next generation Internet protocol selection was not exclusively technical but reflected an international and institutional tension between the dominant Internet establishment versus later Internet entrants poised to change the balance of power and control over the Internet's architecture.

Scholar Arturo Escobar suggests, "The work of institutions is one of the most powerful forces in the creation of the world in which we live. Institutional ethnography is intended to bring to light this sociocultural production."[108] Further Escobar suggests, "The deconstruction of planning leads us to conclude that only by problematizing these hidden practices— that is, by exposing the arbitrariness of policies, habits, and data interpretation and by suggesting other possible readings and outcomes—can the play of power be made explicit in the allegedly neutral deployment of development."[109] Examining IPv6 against its discarded alternatives not only demonstrated institutional tensions but also conflicts among dominant vendors like DEC versus newer entrants like Sun Microsystems, the Internet's grassroots rank and file establishment versus newer institutional formations like the Internet Society, and trusted and familiar insiders versus newer participants.

The selection of IPv6 also occurred outside of the realm of market economics, with the Internet's technical community describing the protocol selection as too complex for markets and suggesting that corporate users, many with ample IP addresses, were not even aware of the presumptive international problem of Internet address space exhaustion. Large American corporations typically had sufficient IP addresses and, at the time, were not demanding a new protocol to expand the Internet address space. If anything, there was market pressure to adopt an OSI rather than TCP/IP-based protocol. The ISO alternative had the political backing of most Western European governments influential technology companies, and users invested in OSI protocols, and was even congruent with OSI

108. Arturo Escobar, *Encountering Development, The Making and Unmaking of the Third World*, Princeton: Princeton University Press, 1995, p. 107.
109. Escobar, p. 123.

directives of the United States. The selection of IPv6, an expansion of the prevailing IPv4 protocol over such a politically sanctioned OSI alternative solidified and extended the position of the Internet's traditional standards-setting establishment as the entity responsible for the Internet's architectural direction.

The IPv6 selection process contained a paradox. The technical community was adamant about eliminating sociological considerations from what they considered a purely technical protocol decision. For example, the IAB had drawn a demarcation between the Internet as a communications system and the Internet as a community of people. Only its architectural constitution could define the Internet. Yet the outcome of the IPng selection process appeared to define the Internet, in part, as the community of *people* who would either retain or gain control of its architecture. A consideration in making architectural decisions related to the next generation Internet protocol seems to have been the retention of the IAB, IESG, IETF institutional structure/people as controlling the Internet's direction rather than relinquishing control to another standards body. Despite the Internet standards community's strategy of eliminating the influence of sociological factors on its architectural decisions, the history of IPv6 indicates that the definition of the Internet, ultimately, includes people.

# 3   Architecting Civil Liberties

This code presents the greatest threat to liberal or libertarian ideals, as well as their greatest promise. We can build, or architect, or code cyberspace to protect values that we believe are fundamental, or we can build, or architect, or code cyberspace to allow those values to disappear. There is no middle ground.[1]

—Lawrence Lessig, *Code and Other Laws of Cyberspace*

Protocols are political. They perform some technical function but can shape online civil liberties in unexpected ways. It is well understood how decisions about encryption protocols must strike a balance between providing individual privacy online and responding to law enforcement and national security needs. Other protocols are not specifically designed to address user privacy but nevertheless have significant privacy implications. This chapter examines how protocol design decisions, including choices made about the final IPv6 specifications, embed the values of standards designers and can serve as alternative forms of public policy not established by legislatures but by Internet designers.

From critical theorist Langdon Winner to legal scholar Larry Lessig, an enormous body of literature examines how technologies embody values and create legal architectures.[2] More recent scholarship has attempted to operationalize theories about values in design into methodologies for externally influencing how technical communities design values into technical architectures. Alan Davidson and Robert Morris have described how Internet standards have particularly complex and important implications for personal privacy, property rights, and public access to knowledge and

1. Lawrence Lessig, *Code and Other Laws of Cyberspace*, New York: Basic Books, 1999, p. 6.
2. See, for example, Langdon Winner, "Do Artifacts Have Politics?" 109 *Daedalus* (Winter 1980), and *The Whale and the Reactor: A Search for Limits in an Age of High*

have accentuated the role of public policy experts and advocates in introducing values considerations into Internet standards design.[3] In practice, methodologies for intervening in standards-setting processes have many economic, technical, and political limitations.

This chapter examines the IETF's decision-making process in finalizing the IPv6 protocol. As Internet engineers tended to the technical details of IPv6 in the years following its selection as the next generation Internet protocol, they grappled with design decisions that would, in effect, establish public policy about user anonymity and privacy. This chapter describes how Internet engineers opted to design some privacy protections into the IPv6 address design. It also examines contemporaneous concerns that privacy advocates raised about IP address privacy, particularly in the European Union. The chapter concludes with an examination of the implications of private institutions establishing public policy, the question of institutional legitimacy, and the issue of how, given technical barriers to participation, the public interest can enter these decisions.

## Values in Protocol Design

The question of whether technologies are fundamentally apolitical or whether they embody values has long been examined by philosophers, critical theorists, and legal scholars alike. This question embeds at least four distinct problems: the possibility of an intrinsic politics embedded within a technological artifact, once developed; the politics and values entering the initial design of the technology; the values reflected in how the technology is actually used; and the question of who judges the intrinsic morality of an artifact or how the artifact is used. This chapter focuses on one narrow aspect of these questions—how values enter, or should enter, Internet protocol design.

Langdon Winner's influential 1980 essay, "Do Artifacts Have Politics?" described two senses in which technologies can embody politics, with politics broadly defined as arrangements of power and authority among humans. In one sense, the design of a specific technology can resolve an

---

*Technology,* Chicago: University of Chicago Press, 1986; see also Lawrence Lessig, *Code and Other Laws of Cyberspace,* New York: Basic Books, 1999.

3. John Morris and Alan Davidson. "Policy Impact Assessments: Considering the Public Interest in Internet Standards Development." *31st Research Conference on Communication, Information and Internet Policy,* August 2003. Accessed at http://www.cdt.org/publications/pia.pdf.

issue within a community, such as the selection of the next generation Internet protocol, inter alia, deciding which community would retain control over the selected protocol. In another sense, Winner cites "inherently political technologies" that coincide with certain types of political relationships.[4] As Andrew Feenberg describes in *Questioning Technology*, "the choice between alternatives ultimately depends neither on technical nor economic efficiency, but on the 'fit' between devices and the interests and beliefs of the various social groups that influence the design process."[5]

Philosophers have also extended discussions about the politics of technology into more specific questions about the *value*s that enter, and that should enter, the formation of information and communication technologies. Helen Nissenbaum has written extensively on this subject, noting that the question is not just one of extending prevailing contextual value beliefs (e.g., about privacy or intellectual property) into information and communication technologies, but of understanding ways in which technologies force a reconceptualization of these values.[6] For example, some scholars view rapidly changing information and communication technologies as reflecting and enlarging democratic values of equal participation, freedom from bias, individual autonomy, and privacy, among other values.

A variety of technical design communities have formed movements to consciously design values into technologies. Noëmi Manders-Huits and Michael Zimmer have described the challenges of what they term "values conscious design," noting that many attempts to self-consciously incorporate values into technological design have concentrated on instrumental norms of safety and user-friendliness rather than on issues of more explicitly "moral import, such as privacy or autonomy."[7] Some of these more instrumentalist-oriented design movements include participatory design communities, advocating that democratic participation in design will more adequately ensure safety and user well-being; and the human–computer interaction (HCI) movement, seeking to improve the usability of technology.

4. Langdon Winner, "Do Artifacts Have Politics" in *The Whale and the Reactor: A Search for Limits in an Age of High Technology*, Chicago: University of Chicago Press, 1986, p. 20.
5. Andrew Feenberg, *Questioning Technology*, London: Routledge, 1999, p. 79.
6. Helen Nissenbaum, "How Computer Systems Embody Values," *IEEE Computer*, March 2001, p. 118.
7. Noëmi Manders-Huits and Michael Zimmer, "Values and Pragmatic Action: The Challenges of Engagement with Technical Design Communities," 2007, p. 4. Manuscript submitted for publication.

Some value-conscious design efforts have attempted to influence technological design to more explicitly embed moral values such as privacy.[8] These so-called value-conscious design frameworks generally recommend a three-part methodology: first identifying the relevant values possibly at play in considering a technological design, second, architecturally translating this value into the design, and finally, evaluating how successfully the technology, once developed, reflects this value in practice.

An example of designing the value of privacy into technical architecture is Howe's and Nissenbaum's TrackMeNot web browser extension, intended to protect user privacy by concealing a user's web search history.[9] The development of TrackMeNot was a response to concerns about the user search query practices of corporations such as AOL, Google, and Yahoo!. The developers of TrackMeNot were concerned about the systematic logging and storing of Internet search queries, as well as the possibility of accidental or intentional release of this data to the public or to third parties. Alternatives for addressing such privacy concerns can take the form of legal, economic, or technical interventions. For example, lawmakers in the United States could choose to extend Fourth Amendment constitutional protections to Internet search data. Rather than such direct government intervention, another alternative would be a laissez-faire approach of allowing autonomous and free markets to select search engines based in part on the desired level of user privacy. One complication is that markets do not have precise knowledge about the search query privacy practices of relevant companies, nor understand that logging and storing of search engine data is occurring at all. Another approach would be the voluntary adoption of best practices for search engine queries on the part of search engine companies. Finally, the TrackMeNot web browser extension is an attempt to use technical architecture to address privacy concerns, allowing individuals the choice of implementing technologies that provide privacy protections. TrackMeNot is downloadable software that runs within

8. See, for example, Mary Flanagan, Daniel Howe, and Helen Nissenbaum, "Values at Play: Design Tradeoffs in Socially-Oriented Game Design," *Conference on Human Factors in Computing Systems*, 751–60, 2005; B. Friedman, D. Howe, and Ed Felten, "Informed Consent in the Mozilla Browser: Implementing Value-Sensitive Design," *Proceedings of the 35th Annual Hawaii International Conference on System Design*, 2002. Accessed at http://www.hicss.hawaii.edu/HICSS_35/HICSSpapers/PDFdocuments/OSPEI01.pdf

9. To download TrackMeNot code, and for information about TrackMeNot, see http://mrl.nyu.edu/~dhowe/trackmenot/.

the Firefox web browser. The software works through obfuscation, issuing periodic random Internet search queries to search engines to mask data and complicate the meaningful tracking of search data on the part of search engine companies.

This technical architecture solution has two limitations. It requires that users understand the possible threats to individual privacy in searching the Internet. If average Internet users are not cognizant of the search query practices of relevant corporations, they will not understand the need, never mind the possibility, of implementing an additional level of privacy via technical architecture. Similarly, it requires not only knowledge but action on the part of users. Individuals must be aware of the existence of Track-MeNot and must be able to access, download, and install the software. Despite any limitations, this type of solution illustrates how designers can embed values into technological architecture.

In this example it is easy to identify the belief systems and values underlying the technical design. Values are also readily identifiable in technical and scientific areas such as norms about human subjects' research, public safety features in technologies, privacy protections built into electronic patient record systems, and human–computer interface features that provide accessibility for the disabled.

In more abstract and concealed areas such as within the Internet's underlying technical protocols, it is more difficult to identify the values that enter, or should enter, technical design. Questions about values in design are much more complicated when applied to technical protocols. Internet protocols are not tangible artifacts like hardware and are not downloadable software code like TrackMeNot. Technical protocols, in general, exist at a much more invisible level. Another complication is that a single hardware or software product can embed numerous protocols. The number of standards required for Internet-based communications has obviously increased as the types of information supported by the Internet have expanded from text to multimedia applications (e.g., video, images, audio) and as the devices for accessing the Internet have become much more diverse (e.g., cell phones, iPhones, Blackberries, laptops). The standards necessary for communicating extend far beyond traditional TCP/IP-related standards such as FTP, SMTP, and HTML. Information exchange can only occur through the use of the basic building blocks of information exchange, such as image formats, video formats, audio formats, and office application formats.

A related complication is that a single Internet device integrates functionality previously provided by multiple devices and thus incorporates

numerous standards established by numerous standards-setting organizations. For example, a single device can provide mobile voice telephony,
web browsing, text messaging, digital imaging, video recording, and other
functions and has the ability to connect to multiple networks like GSM,
Wi-Fi, or a global positioning system (GPS). This type of device has to
integrate hundreds of standards. Individual Internet users are not necessarily even aware of the existence of all these standards, never mind understanding or accounting for ways in which values have entered the
conception and design of such technologies.

Even more difficult than examining ways in which values enter the design
of protocols is the task of normatively proposing methodologies for influencing technical design to proactively reflect certain values. Especially in the
case of Internet protocols, the politics of use of these embedded protocols
can change in different social and political contexts and the question of who
evaluates whether these uses intrinsically reflect "good" or "bad" values is
intractable. Focusing on the values in a protocol design question is itself
complicated. Assessing the values reflected in the development of a polio
vaccine or a technological construct such as a wheelchair, electric chair, or
gas chamber, is different from understanding the values that can enter the
development of intangible and abstract technical instruments like network
protocols. Another complicated dimension is that a significant percentage
of computer users, even those aware of the role of protocols in the technologies they use, are not aware of, never mind involved in, the protocol development process. The question then becomes *whose* values *are* reflected,
*should* be reflected, or realistically *could* be reflected.

The IETF process itself self-consciously expresses certain values. Some
examples of these values include: (1) *universality and competitive openness*—
one objective of developing a standard is for it to become widely used in
the marketplace; (2) *participatory openness* in the standards-setting process;
and (3) the *end-to-end architectural design principle* specifying that intelligence should be located at network end points rather than in medias res.[10]
To elaborate on one of these values, the overall goal of the IETF's standards-
setting process is for the standard to be "widely used and validated in the
marketplace."[11] This may sound obvious to some, but the goal of technical

10. See, generally, Brian Carpenter, ed., "Architectural Principles of the Internet," RFC
1958, June 1996, and J. Saltzer, D. P. Reed, and D. D. Clark, "End-to-End Arguments
in System Design." 2 *ACM Transactions on Computer Systems* 27–288 (November 1984).
11. Susan Harris, ed., "The Tao of IETF—A Novice's Guide to the Internet Engineering Task Force," RFC 3160, August 2001.

standards setting could also be (and often is) to limit implementations based on a standard through intellectual property restrictions such as patents and licensing fees. The goal of limiting availability of the standard is usually to gain market dominance by restricting the variety of products based on the standard. The general rule within the IETF is to use non-patented technology when possible to encourage the maximum implementation of a standard.

This chapter now examines an IPv6 design decision related to user privacy: how Internet engineers identified privacy as a value pertinent to IPv6 address design and embedded this value into design choices.

### Privacy Design Choices

When information is transmitted over the Internet, it is accompanied by a unique address associated with the transmitting device and a unique address associated with the destination device. Messages are routed to the appropriate destination based on the corresponding address. These addresses have historically been software-defined and not associated with any physical architectural component such as a hardware device. Recall that the IPv4 standard used a 32-bit Internet address and that the IPv6 standard expanded this address length to 128 bits. Internet engineers working on the IPv6 specification had to determine how the new IPv6 number would be derived.

In constructing the technical details of how an IPv6 device, such as a personal computer, would generate this 128-bit IPv6 address, one approach originating in the IETF proposed the embedding of a computer's hardware serial number into some IPv6 addresses. This potential incorporation of a physical hardware address within a software-defined IP address would create an environment in which information transmitted over the Internet could potentially be traced to a specific piece of hardware, and therefore possibly traced to a specific computer and an individual's identity and physical location. The following section provides some technical explanation about this design issue.

### Embedding a Hardware Serial Number into an Internet Address

The physical hardware address in question is most easily described as an Ethernet address.[12] This address is distinct from an IP address. To access a local area network (LAN), each computer requires a networking hardware

12. Or other LAN address.

component known as a network interface controller (NIC). Also called a network interface card, network adapter, or simply a network card, the NIC provides the physical interface between a computer and a local medium such as twisted pair cable, fiber optic cable, or free space in the case of wireless LANs. NICs also support an addressing system necessary for exchanging information over a LAN. A NIC used in an Ethernet network is usually called an Ethernet card, which, in the late 1990s context in which Internet engineers were designing IPv6, was typically a circuit board slipped into a computer slot.

Each Ethernet card contains a unique address used to send and receive information over a network. This number, called the media access control (MAC) address, is a unique 6-byte (48-bit) number physically associated with a computer's Ethernet card. To transmit information on an Ethernet LAN from a source computer to a destination computer, the source computer transmits both its address and the destination computer's Ethernet address along with actual information content to be exchanged. In addition to providing addressing functions, the Ethernet card converts information provided by the computer into small groups of bits, called frames. Frames contain the actual information content to be transmitted, along with ancillary information such as the Ethernet addresses of the transmitting and destination computers.

The first three bytes of a 6-byte Ethernet address are called the organizationally unique identifier (OUI), a unique code assigned by the IEEE to a manufacturer of Ethernet cards. The IEEE is responsible for establishing Ethernet standards and has historically allocated 3-byte codes to each Ethernet NIC manufacturer. The manufacturer then assigns a unique number to the remaining three bytes on each Ethernet card it produces. The result is a unique number physically assigned to each Ethernet card. Computing devices read this 48-bit number, but the number physically inscribed on the outside of the Ethernet card is written in hexadecimal as shorthand to make it easier for humans to read. Each hexadecimal character (e.g., "A") represents a 4-bit binary number (e.g., "1010") so a twelve-character hexadecimal number serves as a shorthand notation for a 48-bit Ethernet address. Figure 3.1 shows an example of an Ethernet address, written both in the binary that computers understand and translated into human-readable hexadecimal.

Ethernet addresses were not designed to be used for wide area networking. Because these addresses were used in local geographic contexts, such as within the confines of a building, the association of these hardware serial numbers with the information transmitted over a LAN was never considered a significant privacy issue.

**48-bit Unique Ethernet Address**

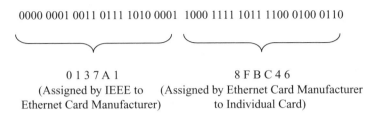

0000 0001 0011 0111 1010 0001  1000 1111 1011 1100 0100 0110

0 1 3 7 A 1                        8 F B C 4 6
(Assigned by IEEE to     (Assigned by Ethernet Card Manufacturer
Ethernet Card Manufacturer)          to Individual Card)

**Unique Ethernet Address (Imprinted in Hexadecimal on Card)**
**0137A18FBC46**

**Figure 3.1**
A unique Ethernet address

In designing the IPv6 address structure, Internet engineers viewed these unique hardware addresses as a possible unique number for computers to use when generating some IPv6 addresses required for sending or receiving information via the Internet, particularly for "stateless address autoconfiguration." IPv6 defines both "stateful" and "stateless" address autoconfiguration. Using stateful autoconfiguration, computing devices obtain an address from a server, which tracks which addresses have been assigned to each computing device.[13] The server is usually called a Dynamic Host Configuration Protocol (DHCP) server.

Stateless autoconfiguration is an approach in which a computer, independent from a server, generates its own address, formed from a combination of a router-provided address prefix associated with a specific network segment and a local number that uniquely identifies a node on the network segment.

Questions about privacy primarily entered the design decisions about stateless address autoconfiguration. Computers form these IP addresses by combining a router-provided network prefix with a locally generated number called the interface identifier. Some engineers proposed that this interface identifier be derived from the IEEE-assigned, globally unique 48-bit address associated with the computer's Ethernet card. Computers would use this unique 48-bit hardware address to generate a 64-bit interface identifier. The primary technical rationale behind incorporating this

13. For more information about stateful and stateless address autoconfiguration, see Susan Thomson and Thomas Narten, "IPv6 Stateless Address Autoconfiguration," RFC 1971, August 1996. Accessed at http://www.tools.ietf.org/html/rfc1971.

hardware number into the IP address was that the number would automatically be globally unique, a requirement for sending or receiving information via the Internet.

### The Privacy Implications of Embedding an Ethernet Address within an IP Address

Embedding a hardware address within a global Internet address would raise three privacy concerns: anonymity, pseudonymity, and location privacy. The use of a hardware serial number within an address is a specific occurrence of a more general case of using any permanent identifier, whether an IP address or other identifier, repeatedly over a prolonged period of time from a single device.

**Anonymity and Pseudonymity Concerns**    Embedding an Ethernet address in an IP address could potentially compromise the anonymity of individual users accessing the Internet. Anonymity, derived from Greek, means literally "without a name." Truly anonymous Internet activity would not disclose the personal identity of the user while browsing the web, posting a blog, and so forth. For example, an anonymous message exchanged between two parties would not include personal identity information about either the message sender or message recipient. Embedding an Ethernet number in an IP address could compromise this anonymity. Each unique Ethernet number is associated with an Ethernet card, which in turn is associated with a computer, which can be potentially linked to the name of the computer's owner. The design alternative of embedding an Ethernet address within an IP address would mean that messages transmitted over the Internet would include this potential personal identifier information. Postings on a discussion board, files downloaded, and websites visited could all be potentially linked to the individual's identity via this unique hardware serial number.

Internet engineers also raised the possibility that the "sniffing" (surveillance) of communications using these unique identifiers could potentially compromise the personal safety of the computer's owner. For example, surveillance of a user's network usage patterns could reveal personal information such as when an individual was normally at home.

It is, however, not inevitable that the personal identity of an individual using a laptop or computer would be readily traceable via an Ethernet card. For example, users can take their computers' Ethernet card from another computer. Nevertheless, the repeated use of the same fixed identifier raises privacy questions related to *pseudonymity*. Even though an individual's personal identification cannot necessarily be linked to the individual's Internet transactions, a fixed pseudonymous identification, via the embedded Ethernet address, is permanently linked to the transactions. In other words, a website will know that the same user is returning. In other cases, associations can be made among otherwise independent and unrelated online transactions. In these cases, the number itself would not necessarily identify a user, but this information in combination with other information, such as an ISP releasing the IP address used by an individual during a specific Internet session, would be sufficient to link an Internet transaction to an individual user. Again, this problem can arise any time a fixed identifier accompanies the exchange of information over the Internet.

**Location Privacy Concerns** Some Internet engineers were particularly concerned about a unique privacy issue accompanying the potential IPv6 address design. Mobile users accessing the Internet from different geographical locations could be tracked based on the unique identifier embedded in the IP address. Within the IPv4 technical approach, mobile users would receive completely new addresses in each location from which they accessed the Internet. The network prefix, the part of the IP address associated with the user's current network location, would be assigned on a per-location basis and be different for every access point. The identifier interface, the individually derived part of the IP address, would also be a new number assigned in every location. The user would have network-level anonymity in accessing the Internet because there would be no fixed identifier that could be linked to the individual user. This variation in numerical identifiers for each mobile user would not be retained under the proposed IPv6 address architecture.

Under the proposed IPv6 address approach, a fixed number would follow a user regardless of geographical location. The network prefix portion of the address would still be assigned based, generally, on the user's geographical Internet access vicinity. This assigned network prefix would vary from location to location and be technically associated with the local router. However, the second part of the address, if derived from a unique hardware identifier, would remain fixed regardless of location. For mobile users accessing the Internet from a laptop or other mobile computing device, this address assignment approach raised location privacy concerns, in addition to anonymity and pseudonymity concerns applicable to all users whether accessing the Internet from a fixed location or from multiple geographical users. If the individual's Internet transactions were intercepted, the user could potentially be tracked. If the individual used a search engine or read a blog from various locations, the user could potentially be tracked. The second part of the Internet address would conceivably provide the identity of the user and the first part of the address would provide the general location of the user.

Using a design in which a fixed portion of the IP address remains constant would mean that this fixed identifier would accompany the user regardless of location. As Internet protocol designers described the problem, "This facilitates the tracking of individuals' devices (and thus potentially users)."[14]

**Not an Entirely New Privacy Concern**    IPv4, in historical context, provided individual users with some degree of user anonymity and location privacy because each 32-bit address was not automatically linked to a particular user, location, or hardware component. In the 1990s, a home user typically connected to the Internet via a dial-up connection. When a user "logged on," meaning initiated an Internet connection, a dial-up ISP such as America Online would assign a temporary address for each online session. The ISP would dynamically assign a different address for the user's next session. The effect of this dynamic address assignment was that addresses were shared among multiple users rather than statically affixed to a single dial-up connection. This dynamic allocation approach did not necessarily mean that the ISP was not tracking which address it assigned to each individual user at any given moment. But it did provide greater pseudonymity privacy when users logged on to a web server, used a search engine, posted a message, or engaged in other Internet activity.

This privacy effect had less to do with IPv4 than with the dominant access method of dial-up, an Internet access method using a modem to establish a temporary connection to an ISP via a traditional telephone line. As "always on" broadband connections started proliferating, this privacy element vanished. With broadband approaches such as cable modem access and digital subscriber line (DSL), some users obtained relatively permanent IP addresses they always used for Internet access. This new privacy issue was not dependent on protocol, and therefore not an IPv4 versus IPv6 issue, but instead an issue of static versus dynamically assigned Internet addresses. In the case of static IP addresses, embedding a hardware serial number within the static address posed little additional risk beyond

14. Thomas Narten and Richard Draves, "Privacy Extensions for Stateless Address Autoconfiguration in IPv6," RFC 3041, January 2001. Accessed at http://www.ietf .org/rfc/rfc3041.txt.

the already existing privacy vulnerabilities. This is not the case for mobile users. Embedding a hardware address within an IP address provides mobile users with an additional location privacy concern, as described earlier.

The problem is also not completely unique to situations using stateless address configuration versus receiving an address from a DHCP server. In theory, the address returned from a DHCP server should change over time, but in practice, the server can return the same address repeatedly.

### Concern about IPv6 Privacy within the IETF

The privacy implications of incorporating a hardware serial number within an Internet address were weighed by Internet engineers working on IPv6, as well as other protocols. Some Internet designers feared that exposing a hardware serial number over a network via protocols would potentially create privacy concerns for users or groups of users. Some believed that a hardware serial number used for addressing in a local area network should never be transmitted onto the global Internet. For example, the 1999 Internet-Draft "Privacy Considerations for the Use of Hardware Serial Numbers in End-to-End Network Protocols" recommended that:

Protocols intended to be used over the global Internet SHOULD NOT depend on the inclusion of hardware serial numbers. Protocols intended to be used only in a local IP-based network, which use hardware serial numbers, SHOULD define a means to keep those serial numbers from escaping into the global Internet.[15]

Other Internet designers believed that if Internet protocol implementations did incorporate numbers derived from hardware elements, users should at least have the option of disabling this element and using an alternative approach. In June 1999, IBM's Tom Narten published an Internet Draft called "Privacy Extensions for Stateless Address Autoconfiguration in IPv6."[16] For cases in which IPv6 addresses are generated via IPv6 stateless address autoconfiguration, in other words, generated without a DHCP server, Narten's Internet-Draft described an optional feature that could generate addresses that changed over time. This Internet-Draft noted an important technical consideration that even when transmissions are

15. See, for example, Keith Moore, Internet-Draft, "Privacy Considerations for the Use of Hardware Serial Numbers in End-to-End Network Protocols," January 26, 1999. Accessed at http://www.tools.ietf.org/html/draft-iesg-serno-privacy-00.
16. Thomas Narten, Internet-Draft, "Privacy Extensions for Stateless Address Autoconfiguration in IPv6," June 1999. Accessed at http://www.ietf.org/proceedings/99jul/I-D/draft-ietf-ipngwg-addrconf-privacy-00.txt.

encrypted over the Internet, the IP addresses contained within packet headers and read by routers are not necessarily encrypted.

## A Public Relations Issue

Months after Internet engineers began addressing the IPv6 privacy design question, an industry columnist raised a red flag about privacy to the public. On October 4, 1999, *InternetWeek* columnist Bill Frezza posted a column entitled "Where's All the Outrage about the IPv6 Privacy Threat?" The column was an inflammatory critique of what he viewed as the IETF's decision to universally embed a user's physical Ethernet address into an IPv6 address. The piece mentioned neither the deliberations about IPv6 privacy within the Internet technical community nor the draft IPv6 privacy document. Frezza warned that every packet sent over the Internet would be linked via an Ethernet card identifier to an individual user, something which Internet engineers were considering under certain conditions (e.g., stateless autoconfiguration) and had been trying to address via a technical design overlay which would engineer privacy into IPv6 addresses.

Frezza's commentary, although containing some historical and technical inaccuracies, or at least omissions, was indicative of the types of strong reactions engendered by the IP privacy question. The commentary was also interesting in that it equally critiqued Internet engineers and privacy advocates, as captured in the following brief excerpts:

It's a conundrum that makes one wonder about the motives of the reigning Internet digerati, who spend much of their time assuring us that they are protecting our interests as they quietly arrogate power in the new world order.

Where are the professional privacy advocates on this issue? Let's start with the Electronic Frontier Foundation (EFF) . . . Go search EFF's site and see if you can find a single word about IPv6 and its privacy problems. The EFF's silence is matched by a similar lack of concern from the Center for Democracy and Technology and the Electronic Privacy Information Center, both of which are usually the first to man the barricades when Big Brother comes knocking.[17]

Shortly after the *InternetWeek* column appeared, Sun Microsystems Engineer Alper Yegin posted a message called "IPv6 Address Privacy" to the IPng mailing list, providing a link to the Frezza column and noting that the author seemed unaware of previous discussions within the Internet

17. Bill Frezza, "Where's All the Outrage about IPv6 Privacy?" 783 *InternetWeek* 43 (October 4, 1999).

standards community and of the existence of an Internet standard draft addressing the issue.[18] The responses of mailing list participants expressed a variety of views, but everyone seemed frustrated. Many of the engineers recognized that such articles were potentially very damaging to IPv6 deployment, as well as to the IETF's reputation. Many focused on the technical inaccuracies of Frezza's piece. Some responded that embedding an Ethernet adapter address from an individual's computer into an IPv6 address is one proposed approach, motivated by the requirement of easy address autoconfiguration, nevertheless admitting that "privacy is a concern, so there's an alternate mechanism being defined."[19]

Others noted that the privacy concern was not unique to IPv6. Increasingly users were attaching to the Internet using relatively permanent IP addresses, essentially enabling tracking of Internet activity by address: "As long as your machine has an address that doesn't change, and you're the only user, and you use it for "surfing," and you don't use a proxy server, you are trackable, even with IPv4. This is no IPv6 problem."[20]

The reaction of the privacy advocates Frezza had criticized was immediate. Shortly after Frezza's column ran, the Electronic Privacy Information Center (EPIC) issued an alert entitled "New Internet Protocol Could Threaten Online Anonymity."[21] EPIC compared the IPv6 address approach to a contemporaneous privacy concern about Intel's Pentium III processor chip, which included a personal serial number. Intel's rationale for embedding a personal serial number within a chip was to prevent hardware theft and software piracy and to serve as a security mechanism to authenticate users' identities during electronic commerce transactions. Privacy advocates denounced the chip's potential for enabling tracking of an individual's Internet activities. The pressure from privacy advocates, which included a threatened boycott, prompted Intel to issue a software patch that would disable the default disclosure of this personal serial number. The Intel

18. See Alper Yegin's posting "IPv6 Address Privacy" to the IETF IPng Working Group mailing list, October 7, 1999, accessed at ftp://playground.sun.com/pub/ipng/ipng-mail-archive/ipng.199910.

19. See Steve Bellovin's posting "Re: Privacy Problems in IPv6" to the IETF IPng Working Group mailing list, October 8, 1999. Accessed at ftp://playground.sun.com/pub/ipng/ipng-mail-archive/ipng.199910.

20. Ignatios Souvatzis, "Re: Privacy Problems in IPv6," posted to the IETF IPng Working Group mailing list, October 8, 1999. Accessed at ftp://playground.sun.com/pub/ipng/ipng-mail-archive/ipng.199910.

21. Electronic Privacy Information Center, EPIC ALERT Volume 6.16, October 12, 1999. Accessed at http://epic.org/alert/EPIC_Alert_6.16.html.

Pentium III quandary involved a dispute related to the value of privacy versus values of intellectual property protection and user authentication.

The same day as EPIC issued its IPv6 privacy alert, the Associated Press (AP) ran a story entitled, "Critics Fear Internet Proposal Could Endanger Users' Privacy." The AP article generally described the IETF's proposal to include a unique hardware identification number within IP addresses and added that the issue "illustrates the danger of the unintended potential consequences from arcane design decisions."[22]

The following day, the BBC News picked up this IPv6 privacy narrative in an article entitled "New Internet Could Carry Privacy Risks."[23] Quoting Marc Rotenberg of EPIC, the BBC story warned about the possibility of Internet sites linking these numbers with an individual's name, address, clothing size, and political preference.

As network engineer Guy Davies summarized on the IPng mailing list, "This is serious because, justifiably or not, people believe the BBC."[24]

Steve Deering of Cisco Systems and Bob Hinden of Nokia, co-chairs of the IETF's Next Generation working group, issued a response to these mounting public concerns about Internet privacy. This "Statement on IPv6 Address Privacy" was, in part, a statement of values and, in part, a technical response to specific claims and concerns about IPv6 address privacy. Deering's and Hinden's statement began with a normative assertion: "The privacy of communication is a major issue in the Internet Engineering Task Force (IETF) and has inspired much of the IETF's recent work on security technology."[25]

The IETF statement then described recent press reports as misleading and inaccurate. While acknowledging that one of many approaches to assigning IPv6 addresses does incorporate the unique hardware serial number in question, the authors noted that not all IPv6 addresses would use this approach. For example, IPv6 devices could use manually assigned IP

22. Ted Bridis, Associated Press, "Critics Fear Internet Could Endanger Users' Privacy," *The Topeka Capital-Journal*, October 12, 1999. Accessed at http://findarticles.com/p/articles/mi_qn4179/is_19991012/ai_n11737132.

23. BBC News, "New Internet Could Carry Privacy Risks," October 13, 1999. Accessed at http://news.bbc.co.uk/2/hi/science/nature/473647.stm.

24. See Guy Davies, "Re: More Misinformation on IPv6," posting to the IETF IPng Working Group Mailing list, October 13, 1999. Accessed at ftp://playground.sun.com/pub/ipng/ipng-mail-archive/ipng.199910.

25. See Steve Deering and Bob Hinden, "Statement on IPv6 Address Privacy," November 6, 1999. Accessed at http://playground.sun.com/ipv6/specs/ipv6-address-privacy.html.

addresses, dynamically assigned temporary addresses, or an IPv6 address in which a random number replaces the hardware number. Deering and Hinden suggested that concerns about privacy within the IETF, long before more recent press concerns, led to the development of an option to include a randomly assigned number within the IP address, but that this design had not yet been fully standardized and thus was not yet published.

In addition to the Internet-Drafts already addressing this question, months before this public debate in the fall of 1999, the published minutes of the IPng working group corroborate that there was concern about privacy, but that some were presciently concerned about the potential negative public relations problem that would ensue over the inclusion of hardware serial numbers within IPv6 addresses.[26]

## Architecting Privacy

The IETF's privacy protection alternative for IPv6 was ultimately published in January 2001 in RFC 3041, "Privacy Extensions for Stateless Address Autoconfiguration in IPv6." Authored by Tom Narten of IBM and Richard Draves of Microsoft Research, the document described a privacy-enhancing technique that could be used for the stateless IPv6 address autoconguration approach in which the address is derived from a unique hardware serial number and without the assistance of a server.

The computer-generated address would be formed from a combination of a router-provided address prefix associated with a specific network segment and a local identifier derived from the hardware serial number to uniquely identify a node on the network segment. As described, the concern was that the constant use of this unique serial number would compromise anonymity, pseudonymity, and location privacy. One possible design alternative to mitigate privacy concerns would have been for implementations to always employ a DHCP server, which would allocate addresses that changed over time. Automatically changing the interface identifier periodically would also have provided greater individual privacy.

The IPng working group crafted an approach to create pseudorandom interface identifiers and temporary addresses using an algorithm they designed for this purpose. The temporary address would not derive from a completely random number generation process, which might result in

26. See, for example, the "Privacy Issues with use of EUI-64 IDs" in the meeting minutes of the IPng Working Group, February 1999. Available at http://playground .sun.com/ipv6/minutes/ipng-minutes-feb99.txt.

two computers generating the same number, but instead would produce a temporary pseudo-random sequence dependent on both the globally unique serial number and a random component. The number would be globally unique because it would derive from the interface identifier and from the history of previously generated addresses, but would be difficult for an external node to reverse engineer to determine the source computer.

Internet engineers had to consider the trade-offs of introducing this privacy feature. For example, changing addresses frequently might affect performance because of the processing time involved in deriving the number. RFC 3041 suggests that the "desires of protecting individual privacy vs. the desire to effectively maintain and debug a network can conflict with each other."[27] Fault management systems sometimes use IP addresses to trace the source of network performance problems, a task complicated in a situation in which a computer's address constantly changes. Other members of the Internet's technical community expressed concerns that this privacy feature would make it technically more difficult to defend against distributed denial of service attacks in which a targeted system becomes disabled because it is flooded with thousands of requests from unwitting computers.[28]

It is also important to note that the privacy options Internet engineers built into IPv6 addressing approaches left many remaining privacy questions. As Internet engineers themselves acknowledged, IPv6 privacy extensions would do little to ameliorate situations in which a static IPv4 address (or IPv6 address) or other constant identifier is used, a situation potentially enabling monitoring of a user's Internet activity. Equally important, it was understood that the privacy extensions would only be effective if implemented, either through a de facto setting in software (e.g., in an operating system) or through action by an end user.

### European Union Privacy Concerns

Privacy concerns about the IPv6 addressing structure surfaced even after the IETF's publication of RFC 3041, "Privacy Extensions for Stateless Address Autoconfiguration in IPv6." In Europe, where privacy norms and

27. RFC 3041, p. 12.
28. See, for example, Francis Dupont and Peeka Savola, "RFC 3041 Considered Harmful," expired Internet draft available at http://www.6net.org/publications/standards/draft-dupont-ipv6-rfc3041harmful-02.txt.

regulations are perhaps the most stringent in the world, IPv6 privacy issues were examined by the Working Party on the Protection of Individuals with Regard to the Processing of Personal Data (Working Party), an independent European advisory committee on data protection and privacy established by the European Parliament.[29] The privacy Working Party cautioned that "Privacy issues raised by the development of the new protocol, IPv6 have not been solved yet."[30]

The Working Party was concerned, in part, because the European Commission had already established action for migrating to IPv6 without consulting with the Working Party about the privacy repercussions of a unique identification number possibly integrated into an IP address. It asserted that IP addresses are *personal* data protected under EU Data Protection Directives 95/46 and 97/66. The objective of the EU data protection directives was to safeguard fundamental human rights and freedoms, particularly the right to privacy recognized in Article 8 of the European Convention on Human Rights.[31] The Working Party essentially concluded that it wished to enter into a dialogue with the IETF. It also invoked how, under EU legislation, access providers and equipment providers have some obligation to both inform users of risks and of implementing privacy techniques as default settings. In summary, the Working Party concluded: "Protocols, products and services should be designed to offer choices for permanent or volatile addresses. The default settings should be on a high level of privacy protection. Since these protocols, products and services are continuously evolving, the working group will have to monitor closely the developments and to call for specific regulation if necessary."[32]

29. The European Parliament established the data protection and privacy working group under Directive 95/46/EC.
30. Article 29 Data Protection Working Party, "Opinion 2/2002 on the use of unique identifiers in telecommunication terminal equipments: the example of IPv6," p. 2. Adopted on May 30, 2002. Accessed at http://ec.europa.eu/justice_home/fsj/privacy/docs/wpdocs/ 2002/wp58_ en.pdf.
31. For more information, see "Directive 95/46/EC of the European Parliament and of the Council of 24 October 1995 on the protection of individuals with regard to the processing of personal data and on the free movement of such data" or the unofficial text available at http://www.cdt.org/privacy/eudirective/EU_Directive_.html. *Official Journal of the European Communities,* November 23, 1995, No L. 281.
32. Article 29 Data Protection Working Party, "Opinion 2/2002 on the use of unique identifiers in telecommunication terminal equipments: the example of IPv6," p. 2. Adopted on May 30, 2002. Accessed at http://ec.europa.eu/justice_home/fsj/privacy/docs/wpdocs/ 2002/wp58_en.pdf., p. 7.

The European Commission's IPv6 Task Force issued a formal response to the Data Protection Working Group's concerns. The IPv6 Task Force acknowledged that using unique identification numbers presents a privacy threat within any communication environment, whether a wireless local area network, a cellular network, or an IPv4 or IPv6 network. But the letter accused the privacy working group of presenting an "unbalanced view" and that, through RFC 3041, IPv6 actually provides greater privacy than IPv4 and recommended that all IPv6 vendors implement RFC 3041. As noted earlier, IPv4 creates privacy questions any time static IPv4 addresses are used. The letter acknowledged that the default IPv6 stateless autoconfiguration approach uses a personal hardware identifier that can be used to trace a user's Internet activity even when the user's device is connected to different networks, but noted that RFC 3041 solves this problem through introducing a random number component. The European Commission IPv6 Task Force not only recommended that IPv6 vendors incorporate RFC 3041 as a default setting in products, but that these products should provide individual users with the ability to enable or disable this privacy feature as desired.[33]

## Protocols and the Public Interest

The privacy implications underlying the design of the IPv6 address structure are an example of how technical standards not only embody values but can serve as a form of public policy determining the extent of individual civil liberties online. Many other protocols directly affect privacy online, especially encryption protocols designed to keep information private during transmission over a network, authentication protocols that keep user identity private, and electronic health care information standards that make decisions about how citizens' health care records are electronically exchanged.

Technological regulations, in the form of protocols, are sometimes more tenacious than traditional regulations. Once adopted, standards permeate technologies made by different manufacturers, and they may endure for long periods of time because of product investments, institutional commitments, and, through network effects, their deep entrenchment in

33. European Commission IPv6 Task Force, "Discussion Document from the European Commission IPv6 Task Force to Article 29 Data Protection Working Group," Version 1.2 (Final), February 17, 2003. Accessed at http://www.ec.ipv6tf .org/PublicDocuments/Article29_v1_2.pdf.

global technology infrastructures. A traditional law can be overturned, and through a political process, change can take effect immediately. Technical standards change in a much different way. A change adopted by a standards institution does not automatically take immediate effect or even mean that an existing protocol will be replaced by a new protocol. Protocols also establish public policy in a much less visible manner. The general public is not necessarily aware of the policies these hidden specification enact. If technical standards make public interest decisions, the questions of *who* sets technical standards and *how* they set them are highly relevant. Power over these standards is not restricted to market power or technological design or efficiency but the ability to make decisions directly impacting the citizens who use technologies. This form of public policy is not established by elected representatives or with public input, but by private actors.

The legitimacy of technical standards setting derives ultimately from expertise. Many of these private actors, while often cognizant of values and concerned with the public interest, are not necessarily in tune with the public interest or trained as such. Those concerned about promoting greater legitimacy in standards setting usually suggest one of three solutions: greater government involvement; direct public participation; or intermediation by advocates.

The possibility of direct government participation in the standards-development process has several complications. One immediate issue is the question of transnational jurisdiction, the determination of whose government has the authority to make policy decisions for a system that transcends national boundaries. Government involvement also introduces tremendous bureaucracy, might not provide the appropriate level of technical expertise in all instances, would be costly, and would introduce a more slow-moving pace than necessary for innovations in information and communication technology. Another complexity is that there are countless standards-setting bodies (all with different procedural norms and membership requirements). Government involvement in the work of all these bodies would be nearly impossible. Certain standards have greater public policy implications than others but it is difficult to predict which standard will be most pressing in this regard. Top-down government intervention would also reverse the traditional approach of Internet standards percolating up from grassroots structures and emanating from working code, a reversal that could have unintended consequences.

Another option for reflecting the public interest is to encourage democratic public involvement in standards setting. Andrew Feenberg summarizes that "technology is power in modern societies, a greater power in

many domains than the political system itself," and consequently democratic standards should be applied to technology like any other political
institution.[34] Direct public participation in standards setting is implausible
for many reasons. There are many barriers to direct public participation in
the standards-setting process. Participation requires a great degree of technical knowledge, time, funding, and awareness. The general public may
not be aware of the existence of protocols, understand the public policy
issues reified in protocols, or even know that standards institutions exist
or why they would participate.

Langdon Winner's notion of the concealed electronic complexity of
information technologies suggests that the public's engagement with
content can convey a misleading sense of control and democratization
even though a complex, underlying technical architecture with public
interest implications exists completely independent of content. Pragmati-
cally, and even if citizens wished to engage directly in technical protocol
design, which standards body would they select; which protocols would
they select; and how would they decide? Protocols originate and develop
before they reach the public sphere.

Some scholars and activists recommend that advocates become involved
in technical standards-setting activities. The Center for Democracy and
Technology (CDT) in Washington, DC, has examined public interest issues
in Internet standards development since before 2000, when it founded its
Internet Standards, Technology, and Policy Project. The starting point of
the CDT's standards work is the assumption that Internet protocols establish public policy in critical areas such as censorship, speech, privacy, and
surveillance.[35] The CDT has participated in several of the IETF's Internet
standards discussions, including the 1999 "Raven Debate" during which
the IETF discussed whether to build wiretapping capability into the Internet's architecture; the Open Pluggable Edge Services (OPES) protocol that
raised issues of data integrity and user privacy; and the GeoPriv working
group seeking to address location privacy issues. The organization was also
involved in the W3C's Platform for Privacy Preferences (P3P)[36] specification
for web privacy.

34. Andrew Feenberg. *Questioning Technology,* London: Routledge, 1999, p. 131.
35. Alan Davidson, John Morris, and Robert Courtney, "Strangers in a Strange Land:
Public Interest Advocacy and Internet Standards," presented at the Telecommunications Policy Research Conference in Alexandria, VA, September 29, 2002, p. 2.
Accessed at http://www.cdt.org/publications/piais.pdf.
36. Lorrie Cranor, "The Role of Privacy Advocates and Data Protection Authorities
in the Design and Deployment of the Platform for Privacy Preferences." *Proceedings*

The CDT has suggested three models for the involvement of public policy advocates in Internet standards development: (1) direct advocate participation within standards design and deliberations, (2) ad hoc presentations and written submissions to standards bodies, and (3) external monitoring of standards bodies. The CDT has also explored more systematic approaches to institutionalize public policy concerns within the Internet standards-setting process. By Internet standards bodies, the CDT primarily is referring to the IETF and the W3C, two of the most prominent Internet standards-setting organizations.

The prospect of interjecting public interest advocates in standards-setting processes has its own limitations. The first problem is one of legitimacy. The involvement of a public-interest advocate from a nonprofit organization contributes no additional legitimacy to the design process. The advocate is not an elected official any more than are Internet engineers involved in protocol design. The participation of someone who understands legal and cultural issues related to privacy can contribute an important perspective, but not one that creates any additional political legitimacy for a private body to establish public policy. The second limitation is an issue of scalability. Participation is resource intensive, requiring considerable technical expertise and usually enormous amounts of time. There are countless organizations setting countless standards. The sheer number of advocates necessary to become involved in all these efforts, even only the activities of a single institution such as the IETF, would be prohibitive. A final consideration is that advocates are often funded by corporations, a relationship potentially influencing their positions and, because these funding sources might not be disclosed, likely resulting in less rather than more transparency.

## Openness as a Value in Protocol Design

The IPv6 privacy features Internet engineers developed are an example of how technical protocols can stand in for law and illustrative of the role of technical institutions, rather than traditional governance structures in establishing public policy in the information society. Not all protocols have the same direct implications to civil liberties as the privacy questions underlying IPv6 address structures, and IP addresses generally. In some

*of the 12th Conference on Computers, Freedom and Privacy*, April 16–19 2002, San Francisco, CA. Accessed at http://www.cfp2002.org/proceedings/proceedings/cranor .pdf.

cases, protocol design might be completely immaterial to the public interest. In other cases, protocols such as specifications underlying electronic voting systems or first responder communication systems can have even more pronounced political implications than the IPv6 privacy design choices. Regardless of what sphere of public interest a standard affects, if a technological specification is of significant relevance to an issue of political consequence, then the character of the processes resulting in its formulation are relevant to democratic values. A core question in regard to such processes is the same question relevant to any decision-making procedures of public import: whose voices and interests are allowed input into the decision? Private institutions made up primarily by individuals working for private industry make most Internet governance decisions, including designing protocols. The previous section described the limitations of alternatives seeking to introduce greater government involvement, direct public participation, or the involvement of advocates in technical protocol development.

But another core question relevant to any decision-making procedures of public importance is by what *procedures* are the decisions weighed. The conditions under which such procedures occur are relevant: if a standard is being developed by a private or voluntary institution, then issues of openness and transparency are critical, such as whether the public can freely access a specification and the records of the proceedings concerning its adoption and modification. Whereas the design of technical standards can have significant effects on public and individual issues such as privacy, access, speech, and government accountability and whereas this form of public policy is primarily set by private actors and not by governments, one source of legitimacy is through transparency, transparency about *what* is being done, *how* it's being done and *who* is doing it.

The IETF makes its mailing list deliberations, conference proceedings, meeting minutes, and draft standards publicly available. If it did not provide this degree of institutional transparency, privacy advocates would probably not have been aware of the privacy design choices at hand during the design phase of IPv6. Not all standards-setting organizations provide this transparency, but it is a condition of public importance when institutions weigh decisions about protocols that have political implications. Transparency is not a single characteristic but a principle that translates into disclosure, recordation, and open document availability in numerous areas. The following are ten areas of possible transparency in protocol development:

1. Disclosure of organizational affiliation  Do individual participants disclose their organizational affiliations?

2. Disclosure of funding sources  Who is funding the standards work?

3. Disclosure of membership  Who are the members of the standards institution?

4. Disclosure of intellectual property  Is there ex ante disclosure of standards-based intellectual property?

5. Well-defined procedures  Is the development process well defined and publicly available?

6. Well-defined appeals process  Is there confusion about how appeals processes work and is this information publicly available?

7. Record of public dissent  Is there a process for recording dissent and making this information part of the public record?

8. Publicly available procedural records  Are meeting minutes and electronic discussions part of the public record?

9. Recordation  Are deliberations recorded and made publicly available?

10. Public availability of standard  Are drafts and final standards published and freely available?

These transparency characteristics, when exhibited by standards-setting organizations, can enhance the legitimacy of private institutions in making public decisions. Another aspect of legitimacy is participatory openness: whether anyone can openly and freely participate in debates and deliberations about protocol characteristics. This book has described some effective barriers to participation, such as technical expertise, in IETF working groups. There are also legitimating implications of society viewing the Internet's architecture as democratized because of this openness. Nevertheless, the IETF working group discussing the privacy implications of IPv6 addressing techniques at least provided the possibility of public involvement and made deliberations public enough to alert privacy advocates. Unlike these IETF processes, some standards development processes are closed, require fee-based membership, exclude nonmembers, disallow individual citizen participation, and provide no avenues for public participation or oversight. Such barriers to broad and roughly equal participation and public input are clearly at odds with contemporary understandings of legitimacy and transparency that democratic publics expect with regard to public policy.

Participatory openness and transparency help legitimate the public policy responsibilities of private standards institutions, but they are not principles that address all concerns. First, these characteristics only apply to standards development, not standards implementation or adoption. The existence of IPv6 privacy features does not necessitate the implementation of these features by companies developing IPv6-compliant software. It also has no bearing on whether Internet service providers or end users implement IPv6 or associated privacy extensions. Furthermore, concerns remain about Internet address privacy, whether IPv6 or IPv4 and whether using static or dynamic addressing. Because every information exchange contains the sender's Internet address and because websites may collect IP addresses associated with each transaction, privacy advocates often raise questions about whether this unique identifier is personal information directly linked to an individual user and whether this information deserves privacy protections.

But this episode is a reminder that some of the most critical Internet governance questions concern individual civil liberties and that design decisions can present an opportunity to advance libertarian and democratic values or to contain these values. IPv6 privacy design implications and value-conscious design choices reinforce the notion that Internet architecture and virtual resources cannot be understood only through the lens of technical efficiency, scarcity, or economic competition but as an embodiment of human values with social and cultural effects.

# 4   The Politics of Protocol Adoption

As a cultural enterprise, science, like religion or art, ... while differentiated from politics, can be deployed and adapted as elements of particular political worlds.[1]

—Yaron Ezrahi, *The Descent of Icarus: Science and the Transformation of Contemporary Democracy*

The most striking aspect of the evolution toward a new Internet protocol is the disconnect between promises of imminent migration versus the realities of negligible IPv6 deployment. This chapter shifts attention from IPv6 development within the Internet's standards-setting community to the topic of IPv6 adoption. The IETF completed the core IPv6 specifications in 1998.[2] Beginning in 2000, governments in China, Japan, the European Union, Korea, and India viewed IPv6 as a national priority and inaugurated policies to rapidly drive deployment. The United States, with a dominant Internet industry and ample addresses, remained relatively disinterested in IPv6 until the Department of Defense, in 2003, endorsed the protocol as a potential apparatus in the post–September 11 war on terrorism. IPv6 advocates also promoted the standard as a mechanism for global democratic reform, third world development, and the eradication of poverty. Others warned that US inaction on IPv6 threatened American competitiveness and jobs relative to countries like China and India with aggressive IPv6 strategies.

Despite a decade of expectations about imminent global conversion to IPv6, the real world situation is that IPv6 deployment has been extremely slow. Most of the IPv6 implementations that have occurred have deployed

1. Yaron Ezrahi, *The Descent of Icarus: Science and the Transformation of Contemporary Democracy*, Cambridge: Harvard University Press, 1990, p. 1.
2. For the formal IPv6 draft standard document, see Steven Deering and Robert Hinden, "Internet Protocol, Version 6 (IPv6) Specification," RFC 2460, December 1998.

a dual protocol technical strategy of using both IPv4 and IPv6 protocols, a technique that contravenes the original IPv6 objective of addressing IPv4 address scarcity. This chapter begins by describing the progression of national IPv6 policies and IPv6 advocacy within a variety of political and economic contexts, exploring possible intersections between IPv6 decisions and socioeconomic and political order. The chapter examines how IPv6 adoption plans have not translated into commensurate implementations, concluding with an examination of IPv6 transition struggles and the prospects for the Internet ever upgrading to IPv6.

## The Lost Decade and the e-Japan Strategy

Back in 2000 the newly elected prime minister of Japan, Yoshiro Mori, introduced an e-Japan program establishing a 2005 deadline for upgrading every Japanese business and public sector computing device to IPv6. Mori had commissioned his administration, the "Cabinet for the Rebirth of Japan,"[3] to prioritize economic recovery in the wake of long-term stagnation often designated Japan's lost decade.[4] Rising stock and land prices had dominated the late 1980s, with capital gains on these assets exceeding Japan's gross domestic product (GDP) by 40 percent.[5] The government sought to contain speculative investment through a series of interest rate increases and real estate lending ceilings, resulting eventually in real estate and stock market declines and a 61 percent drop in the Nikkei 225 average between January 1990 and January 1999.[6] Although the Japanese economy had begun to rebound when Prime Minister Mori assumed office, Japan had only just weathered a decade-long recession characterized by economic stagnation and high unemployment.[7] The Japanese people were also anticipating the advent of the new millennium, which they celebrated

3. Yoshiro Mori, "Policy Speech by Prime Minister Yoshiro Mori the 147th Session of the Diet," 7 April 2000. Accessed at http://www.kantei.go.jp/foreign/souri/mori/2000/0407policy.html.

4. Yoshiro Mori, "Shaping Japan, Shaping a Global Future—A Special Message from Yoshiro Mori." Accessed at http://www.kantei.go.jp/foreign/souri/mori/2001/0127davos_e.html on April 16, 2003.

5. *The Economist Intelligence Unit*, "Country Profile Japan 2000—Economic Performance," March 14, 2000.

6. The Nikkei index closed at 37,189 on January 31, 1990, and closed at 14,499.25 on January 29, 1999, a decline of 61 percent.

7. Robert M. Uriu, "Japan in 1999: Ending the Century on an Uncertain Note," 40 *A Survey of Asia in 1999* 143 (January–February 2000).

on January 1, 2001. In contrast to Japan's arduous economic circumstances throughout the 1990s, the prime minister believed the Internet had created positive structural changes in other countries, had engendered productivity improvements, and had inaugurated entirely new industries, especially in the United States.

Within this context the prime minister delivered his first *Session of the Diet*, a constitutionally mandated address to elected representatives in Japan's legislative parliament. Mori selected the promotion of science and technology as his administration's policy cornerstone and envisioned "economic development that capitalizes on the explosive force of the IT Revolution."[8] The prime minister introduced a structural program for the "rebirth of Japan" containing five pillars: the rebirth of the economy, the rebirth of social security, the rebirth of education, the rebirth of government, and the rebirth of foreign policy. The prime minister suggested that economic resurgence was a foremost priority and believed information technology represented a critical ingredient in achieving all his pillar priorities. Information technology would represent the "major key to ensuring the prosperity of Japan in the twenty-first century."[9] Mori announced the establishment of an Office of Information Technology within the Cabinet Secretariat and established a deadline of five years within which Japan would become a leader in information and communications technologies.[10]

Mori also established an IT Strategy Headquarters within the Japanese cabinet, tasked with transforming Japan into a global information technology leader and comprising senior administration officials, including the Minister of Justice, the Minister of Finance, and the Minister of Foreign Affairs.[11] The Cabinet directive establishing the IT Strategy Headquarters

8. "Policy Speech by Prime Minister Yoshiro Mori the 147th Session of the Diet," April 7, 2000. Accessed at http://www.kantei.go.jp/foreign/souri/mori/2000/0407policy.html.

9. Yoshiro Mori, "Statement by Prime Minister Yoshiro Mori at the Eleventh Joint Meeting of the Advanced Information and Telecommunications Society Promotion Headquarters and Their Advisory Council," May 19, 2000. Accessed at http://www.kantei.go.jp/foreign/souri/mori/2000/0519statement-it-html.

10. Yoshiro Mori, "Policy Speech by Prime Minister Yoshiro to the 149th Session of the Diet," July 28, 2000. Accessed at http://www.kantei.go.jp/foreign/souri/mori/2000/0728policy.html.

11. Japanese Cabinet Directive, "Establishment of the IT Strategy Headquarters," July 7, 2000. Accessed at http://www.kantei.go.jp/foreign/it/council/establishment_it.html on April 17, 2003.

also installed an "IT Strategy Council" of industry and academic experts to serve in an advisory capacity. The majority of Strategy Council members represented large Japanese technology corporations. Nobuyuki Idei, chairman and CEO of Sony Corporation, chaired the Council, which also included presidents and CEOs from major Japanese corporations such as NEC Corporation, Fujitsu Research Institute, Nippon Telegraph and Telephone (NTT) Corporation, and professors from several of Japan's universities.[12]

The IT Strategy Council and its corporate membership would play a central role in establishing Japan's technical policy directions. Four months after its inception the Council published its basic IT strategy recommendations for Japan. The Council's strategy contained some blanket assumptions about the significance of information technology in society, the position of Japan in the world IT market, and the causes of Japan's shortcomings. The Council asserted that a worldwide IT revolution was "beginning to bring about a historic transformation of society, much like the Industrial Revolution did from the eighteenth century in the United Kingdom" but that Japan's "backwardness" was precluding Japan from embracing this revolution.[13] By backwardness, the Council suggested Japan trailed the United States, Europe, and other Asia-Pacific countries in information technology usage in business and government and that this sluggishness might create an irreparable competitive disadvantage. The Council's causative attribution of this latency ignored Japan's decade-long economic stagnation, the historical circumstances of Internet technologies emanating originally from the United States, or cultural conditions within Japan. Instead, the Council attributed Japan's competitive disadvantage to a single circumstance. Excessive government regulations, telecommunications fees, and restrictions on the technology industry were responsible for Japan's predicament. The solution to Japan's economic indolence in information technology was the implementation of institutional reforms enabling "free and fair competition."[14] The first of four policy priorities the Council recommended was the promotion of a high-speed[15] network infrastructure accompanied by a shift

12. The complete member list of Japan's IT Strategy Council is included in the Japanese Government's IT Strategy Council announcement. Accessed at http://www.kantei.go/jp/foreign/it/council/council_it.html on April 15, 2003.
13. Japan IT Strategy Council, "Basic IT Strategy," November 27, 2000, Accessed at http://www.kantei.go.jp/foreign/it/council/basic_it.html.
14. Ibid.
15. The Japanese IT Strategy Council's definition of high speed in 2000 was 30 to 100 Mbps (megabits per second).

from regulations-oriented to competition-promoting government attitudes toward the telecommunications industry.

As part of achieving its top priority of a high-speed network infrastructure and accompanying policies, the Council recommended the IPv6 standard. IPv6 was the only standard or even technology mentioned by name in the recommendations and the Council cited the need for more Internet addresses, enhanced security, and requirements to connect wireless devices and home appliances to the Internet as justifications for implementing IPv6. The IT Strategy Council's recommendations lacked reflexivity somewhat in that, on one hand, they denounced competition-stifling governmental dictates as the causative factor in economic stagnation but, on the other, recommended a governmental dictate for industrywide adoption of a single technology, IPv6.

The decision distinguishing IPv6 as a specific technological direction for Japan directly corresponded with technical strategies of the corporations represented on the IT Strategy Council. Some of the Council's participants manufactured consumer electronic devices, lucrative gaming products, or home appliances, and were pursuing a strategy of network-enabling products through embedding of IPv6 addresses. These manufacturers, by 2000, had adopted strategies of producing nothing without an embedded network interface. For example, Sony Corporation envisioned a "broadband network society" in which every television, computing device, telephone, appliance, and gaming product, including its profitable Playstation 2, would possess its own unique IPv6 address.[16]

Japan's IT Strategy Council also included representatives of network service providers and network equipment vendors, corporations with their own IPv6 strategies. In 2000, Japan's market leaders in networking products and services introduced a flurry of new IPv6 product and service offerings. Japanese network service provider, NTT Communications, had already announced the availability of its first IPv6-based Internet service and had trial customers.[17] Nokia announced the availability of an IPv6 service as part of its GPRS (General Packet Radio Service) network. Nokia's rationale for introducing IPv6 services included what it considered constraints on available IPv4 addresses and perceptions of greater security and quality of service in IPv6.[18] Another major IPv6 product announcement was Hitachi's

16. Sony Annual Report 2001, Year Ended March 31, 2001.
17. NTT Press Release, "NTT Multimedia Communications Laboratories Announces First Commercially Available IPv6 IX," March 13, 2000.
18. Nokia Press Release, "Nokia Announces the World's First IPv6 Enabled GPRS Network," November 21, 2000.

expansion of IPv6 support to its entire line of Gigabit speed routers, the GR2000 product family.[19] Hitachi had already included some IPv6 support in its router products dating back to 1997 and believed the world would run out of IPv4 addresses by the year 2001.[20] Japan's NEC and Fujitsu similarly offered new router products incorporating IPv6. In the preceding year US-based router manufacturer, Cisco Systems, dominated the router market with an estimated 77 percent market share.[21] Nortel Networks and 3Com were the number two and three router vendors, with roughly 8 and 3 percent of the worldwide router market. Japanese router vendors, whose market share barely registered relative to these other equipment suppliers, were seeking ways to competitively differentiate, or at least competitively maintain, their product lines and considered IPv6 support one possibility.

Many Japanese corporations associated with the IT Strategy Council also had a history of IPv6 development and testing through participation in WIDE Project, a Japanese Internet research consortium. WIDE Project, short for Widely Integrated Distributed Environment, formed an IPv6 working group in 1995 to address the prospect of IP address space exhaustion and examine the possibility of transitioning to the new protocol. In 1996, WIDE's IPv6 test bed, 6Bone, forwarded its first IPv6 packets. This experimentation preceded the IETF's formalization of the core IPv6 specifications. In 1998 WIDE Project members launched KAME Project, a research effort designed to combine numerous IPv6 software implementations into a single IPv6 software stack integrated into the BSD (Berkeley Software Distribution) operating system.[22] In other words, project members worked to develop free IPv6 software code for variants of BSD. Participants in KAME, (the Japanese word for "turtle") funded their involvement, and most of the core project researchers worked for Japanese technology companies including Fujitsu, Hitachi, Toshiba, Internet Initiative Japan, and NET Corporation. The corporate members of the IT Strategy Council establishing Japan's IT policies were already involved in IPv6 development, had

19. Hitachi News Release, "Hitachi GR2000 Router Supports IPv6," November 29, 2000.
20. Ibid.
21. According to *InternetWeek's* By the Numbers Archive, "Worldwide Router Market Share," citing Dataquest statistics, June 23, 1999.
22. Jun-ichiro itojun Hagino, "Implementing IPv6: Experiences at KAME Project," *Applications and the Internet Workshop*, Symposium Proceedings, January 2003, p. 218.

expressed concern about possible IPv4 addresses shortages, and had an economic stake in IPv6 through the prospect of becoming more competitive with dominant Internet software and hardware companies and service providers.

Two months prior to the Council's official publication of Japan's IT strategy, the prime minister delivered a policy speech in which he discussed social issues like educational reform, social security, and foreign policy, but first addressed a topic he called "The IT Revolution as a National Movement."[23] Reflecting the Council's strategic recommendations, IPv6 was the only specific technology the prime minister mentioned in his address to Japan's joint legislative body. The prime minister promised:

We shall also aim to provide a telling international contribution to the development of the Internet through research and development of state-of-the-art Internet technologies and active participation in resolving global Internet issues in such areas as IP version 6.[24]

The mention of such a specific technical protocol by a prime minister was highly unusual, as was his rhetorical grouping of IPv6 with such issues as foreign policy and educational reform.

Following the prime minister's mandate for Japan to pursue IPv6 as part of a national strategy, the IT Strategy Headquarters formally issued its *e-Japan Strategy* (January 2001). The *e-Japan Strategy* reiterated verbatim the IT Strategy Council's recommendations with the addition of specified deadlines for achieving priorities. The *e-Japan Strategy*'s overall objective was to elevate Japan to a global IT leader within five years. Achieving this objective would require Japan transitioning to an IPv6 Internet environment by 2005.[25] The government's comprehensive mandate included myriad strategies to drive adoption: spending eight billion yen on IPv6 research and development in 2001, offering tax incentive programs to IPv6 developers and providers, and instituting educational campaigns to encourage migration.[26] The Japanese government also launched an IPv6 advocacy group called the IPv6 Promotion Council of Japan.

23. Yoshiro Mori, "Policy Speech by Prime Minister Yoshiro Mori to the 150th Session of the Diet," September 21, 2000.
24. Ibid.
25. Specified in the e-Japan Priority Policy program, Policy 2, March 20, 2001. Accessed at http://www.kantei.go.jp/foreign/it/network/priority/slike4.html on April 15, 2003.
26. According to the presentation by the Co-chair of the IPv6 Promotion Council of Japan and board member of the IPv6 Forum, Takashi Arano, at the Shangai ICANN meeting, October 28, 2002.

The *e-Japan Strategy* and especially the prime minister's personal endorsement of IPv6 raised awareness of IPv6 among the Japanese people, but not
everyone agreed that a top-down mandate to drive IPv6 adoption was
prudent or necessary. Nobuo Ikeda, a senior fellow at the Research Institute
of Economy, Trade, and Industry (REITI) and Professor Hajime Yamada
issued a technical bulletin challenging many of the Japanese government's
assumptions about IPv6.[27] They challenged the notion that IPv4 addresses
were critically scarce and disputed the e-Japan program's assertion that
IPv6 provided novel functionality such as improved security or privacy.
For example, they noted the IP security standard, IPsec, could accompany
either IPv4 or IPv6, although it was often cited as a reason for upgrading
to IPv6. Ikeda and Yamada especially challenged the merits of Japanese
government mandates versus a public, national debate, suggesting that
"debate on these fundamental issues concerning IPv6 has been neglected
in Japan, and instead the nationalistic argument that the United States
enjoyed an exclusive victory with IPv4, so Japan should strike back with
IPv6 is being raised."[28] The authors suggested the top-down mandate from
the Japanese government reversed the historical trajectory under which
the Internet had progressed and also raised the question of whether the
rest of the world would even transition to IPv6.

## European Union Internet Strategy

Contemporaneous to Japan's sweeping mandate, the European Union
announced a pan-European IPv6 upgrade. This emphasis on homogeniza-
tion of technology standards accompanied the integration of monetary
standards under the Euro, and reflected the general zeitgeist of European
unification objectives. In March 2000 European Union leaders convened
in Lisbon, Portugal, to formally inaugurate a litany of national and pan-
European reforms. This meeting of the European Council in Lisbon established a sweeping objective for the European Union to overtake US
dominance of the IT market and "become the most competitive and
dynamic knowledge-based economy in the world, capable of sustainable
economic growth with more and better jobs and greater social cohesion."[29]

27. Nobuo Ikeda and Hajime Yamada, "Is IPv6 Necessary?" Technology Bulletin:
Series 2, *GLOCOM Platform from Japan*, February 27, 2002.

28. Ibid.

29. Lisbon European Council, *Presidency Conclusions*, March 23–24, 2000. Accessed
at http://ue.eu.int/ueDocs/cms_Data/docs/pressData/en/ec/00100-r1.en0.htm.

The Council cited concerns about Europe's unemployment rate and identi-fied telecommunications and the Internet as an underdeveloped sector poised to strengthen the region economically. The Council posited that increased understanding and diffusion of Internet technologies would increase employment rates and enable the European Union to "catch up with its competitors" in these areas.[30] One outcome of the Lisbon summit was a call for an "eEurope Action Plan."

The European Council and the Commission of the European Com-munities later issued a 2000 eEurope Action Plan identifying areas in which cross-European action might advance the Lisbon objectives of developing a "new" network-enabled knowledge-based economic structure capable of improving European global competitiveness. "Rapid deploy-ment and use of IPv6"[31] ranked among specific action items for achieving this vision.

The EU 2000 IPv6 announcement cited "the need for vastly increased Internet IP addresses"[32] as a justification for a comprehensive IPv6 conver-sion. An unquestioned assumption was that the IPv4 address space would become "critically scarce by 2005."[33] A significant consideration in the EU decision to advance IPv6 included the planned deployment of third-generation (3G) wireless networking, itself a technology standardization effort enmeshed in a complex array of economic and political circum-stances. At the onset of the twenty-first century, more than 60 percent of Europeans used mobile telephones primarily through GSM (Global System for Mobile communications) service subscriptions, also called 2G, or second-generation wireless.[34] GSM service offered a digital upgrade from what technologists would retrospectively label "first-generation" analog

30. Ibid.

31. eEurope Action Plan prepared by the Council of the European Union and the European Commission for the Feira European Council, Brussels, Belgium, June 14, 2000, p. 6. Accessed at http://europa.eu.int/information_society/eeurope_en.pdf on November 11, 2002.

32. Ibid.

33. Commission of the European Communities, Communication from the Com-mission to the Council and the European Parliament, "Next Generation Internet-Priorities for Action in Migrating to the new Internet Protocol IPv6," Brussels, Belgium, February 21, 2002. Accessed at http://www.ec.ipv6tf.org/PublicDocuments/com2002_0096en01.pdf on November 20, 2002.

34. Commission of the European Communities, Communication from the Com-mission to the Council, the European Parliament, the Economic and Social Com-mittee and the Committee of the Regions, "The Introduction of Third Generation

mobile technology. The European Union, trailing the United States in Internet software and hardware markets, recognized the anticipated convergence between Internet applications and mobile telephony and believed it could leverage its mobile phone diffusion and expertise to globally dominate markets for high-speed mobile Internet services. Consequently, they decided to adopt the International Telecommunications Union's (ITU) recommended family of high-speed, digital, wireless standards known as 3G. The European Parliament established legislation dictating how member states would grant licenses for the 3G frequency spectrum.[35] By March 2001 purchases of 3G licenses, primarily through spectrum auctions, amounted to more than 130 billion euros.[36] Telecommunications operators intending to eventually sell 3G services incurred these spectrum costs, which excluded the enormous expenditures of deploying completely new wireless communications infrastructures. The auctions only sold rights to the invisible resource of airwaves. Telecommunications operators raised massive capital through financial markets and debt instruments to acquire spectrum. The European Commission recognized the great risks inherent in massive radio spectrum expenditures, including delays in availability of 3G handsets, without which 3G services would be useless, and delays in 3G network equipment components.[37]

The European Commission also linked the estimated success of 3G systems to another invisible resource, IP addresses. Providing Internet connectivity via a 3G wireless platform would require an IP address, which the European Union considered in scarce supply. A 2001 European Commission Report on the introduction of 3G mobile communications warned:

---

Mobile Communications in the European Union: State of Play and the Way Forward," Brussels, Belgium, March 20, 2001, p. 4. Accessed at http://europa.eu.int/ISPO/ infosoc/telecompolicy/en/com2001-141en.pdf on November 20, 2002.

35. Directive No 13/1997/EC of the European Parliament and of the Council, April 10, 1997. Accessed at http://europa.eu.int/eur-lex/lex/LexUriServ/LexUriServ .do?uri=CELEX: 52001DC0141:EN:HTML on November 21, 2002.

36. Commission of the European Communities, Communication from the Commission to the Council, the European Parliament, the Economic and Social Committee and the Committee of the Regions, "The Introduction of Third Generation Mobile Communications in the European Union: State of Play and the Way Forward," Brussels, Belgium, March 20, 2001, p. 6. Accessed at http://europa.eu.int/ISPO/ infosoc/telecompolicy/en/com2001-141en.pdf on November 20, 2002.

37. Ibid.

The current implementation of the Internet Protocol (version 4, IPv4) is considered to limit the full deployment of 3G services in the long run. The proposed new IP version (IPv6) would overcome this addressing shortage and enable additional features, such as guaranteed quality of service and security. . . . Any delay in the transition to all-IPv6 networks, which will require several years of effort, risks hindering the deployment of these advanced 3G service features at a later state.[38]

European Commission policies linked IPv6 expertise and deployment with economic opportunities in 3G services and emerging Internet technologies, with achieving its objective of the European Union becoming a competitive knowledge-based economy, and with reducing unemployment.

In 2002 both European and Asian leaders, sometimes working in consort, elevated the need for IPv6 with such issues as weapons of mass destruction disarmament and eradicating poverty. The 2002 annual Japan–European Union Summit, held in Tokyo, addressed a number of joint political objectives. The first objective addressed promotion of peace and security, including weapons disarmament and reconstruction assistance to Afghanistan. The second objective addressed broad prescriptions about fighting poverty, strengthening the international monetary system, and regulatory reform, but also contained one esoteric prescription: a call for "Expert meetings on the fourth [sic] generation mobile telecommunications system and IPv6."[39] The joint statement came from the prime minister of Japan and the prime minister of Denmark in his capacity as president of the European Council, another example of European leaders singling out IPv6 over numerous other technologies and aligning expectations of IPv6 with specific political and economic objectives.

## IPv6 Momentum in Asia

The Korean government similarly announced an objective of rapidly developing IPv6 networks and products in February 2001, when Korea's Ministry of Information and Communication issued a strategic blueprint termed the IT839 Strategy. Between 2000 and 2001, information technology exports, particularly of semiconductor products, experienced a precipitous

38. Ibid., p. 8.
39. The 11th summit between Japan and the European Union. Joint press statement of Junichiro Koizumi, prime minister of Japan, Anders Fogh Rasmussen, prime minister of Denmark, and Romano Prodi, president of the European Commission, Tokyo, Japan, July 8, 2002.

decline of 21 percent.[40] Emphasizing that information technology products comprised 30 percent of Korean exports, the IT Strategy's objective was to "open the era of $20,000 GDP per capita."[41] The nomenclature 8-3-9 indicated that Korea would promote eight new services (e.g., radio frequency identification sensor technologies), three infrastructures, and nine new growth engines (e.g., next generation mobile communications). Korea's strategy cited the economic potential of serving emerging technology markets like wireless broadband and Internet telephony (e.g., VoIP) and itemized three necessary infrastructural developments to achieve its goals: broadband convergence networks providing high-speed multimedia access, ubiquitous sensor networks to improve the management and distribution of food and products, and IPv6.

The Korean strategy embraced the assumption that IPv4 addresses would become depleted by 2006 but emphasized the overall objective of becoming "an Internet powerhouse by promoting IPv6."[42] The Ministry of Information and Communication initially committed $150 million dollars for pilot projects and funding of Korean manufactured routers supporting IPv6. The Ministry also established an IPv6 Strategic Council to promote collaboration among industry, government, academics, and research institutions. The Korean government expected significant returns on its IPv6 investment: "The successful promotion of IPv6 will create 8.6 trillion *won* in production and 53,000 new jobs." Considering that IPv6 was a networking standard for routing and addressing and not an actual application sold to end users, South Korea expected it would sell IPv6 equipment. Relative to the worldwide router market in 2001, the estimate of selling 8.6 trillion *won* (at the time, approximately 8 billion dollars) worth of IPv6 products appeared extremely optimistic.

Japan, the European Union, and Korea were frontrunners in the early promotion of IPv6 products, services, and adoption. India and China, the two countries with the largest potential Internet services user markets, later issued similar sweeping mandates. In 2004 India's Minister of Communications and Information Technology included the goal of national migration to IPv6 by 2006 in his "Ten Point Agenda" for promoting economic development in

40. From the statistics of Korea's Ministry of Information and Communication. Accessed at http://eng.mic.go.kr on January 29, 2006.
41. Ministry of Information and Communication, Republic of Korea, "The Road to $20,000 GDP/capita, IT839 Strategy." Accessed at www.mic.go.kr on January 28, 2006.
42. Ibid.

information technology in India.[43] The Indian government established 2006 as the target for all of India's Internet service providers to upgrade to IPv6.

China began testing IPv6 in 1998 by developing the China Education and Research Network (CERNET) IPv6 test bed. Established with federal government funding and Chinese Ministry of Education oversight, CERNET would eventually interconnect twenty-five universities in twenty cities.[44] In 2002, China entered into a joint initiative with Japan to undertake an IPv6 test bed called the Sino–Japan IPv6 trial network, IPv6-CJ. Also in 2002, the Chinese government established a "National 863 Program, Comprehensive Experimental Environment for New Generation Internet Technology," and an objective of the Chinese IPv6 strategy was to earmark significant funding to support domestic router development.[45]

In 2003 China formally announced its national IPv6 strategy to develop a nationwide IPv6 backbone, the China Next Generation Internet (CNGI).[46] All five of China's national service providers—China Telecom, Unicom, Netcom, China Mobile, and China Railcom—along with CERNET would participate in the national CNGI IPv6 network. In addition to concerns about projected Internet address scarcity, the government sought to encourage China's router manufacturers to develop IPv6-enabled products for use in domestic networks and to potentially gain market share in the global router market dominated by American router manufacturers such as Cisco Systems and Juniper Networks. China's Next Generation Internet Project was a government-sponsored IPv6 initiative designed primarily to propel China's reputation both as a high-tech producer and user of Internet technologies and also to gain first-mover advantage economically. Seeking to propel its reputation as a technologically advanced nation and leader in IPv6, China received a great deal of press when it announced that it would showcase its IPv6 capability in the context of the 2008 Beijing Olympics. China made the official Olympic website accessible via IPv6

43. Thiru Dayanidhi Maran, "Ten Point Agenda Declared by Hon'ble Minister of Communications and Information Technology," May 26, 2004. Accessed at http://www.dotindia.com on June 12, 2005.
44. Chinese Ministry of Education, "Chronicle of CERNET, 1999–2003." Accessed at http://www.edu.cn/20041125/3122220.shtml on December 20, 2005.
45. Hua Ning, Chief Engineer, Beijing Internet Institute, "IPv6 Test-bed Networks and R&D in China," *Proceedings of the 2004 International Symposium on Applications and the Internet Workshops*, IEEE Computer Society, 2004.
46. Jie An and Jianping Wu, of CERNET/Tsinghua University, "CNGI and CERNET2 Updates," November 2, 2005. Accessed at http://cans2005.cstnet.cn/down/1102/A/morning/200501101-wjp-aj-CERNET2_1A.pdf on December 20, 2005.

networks, but the use of IPv6 within the context of a "digital Olympics" was not necessarily geared toward traditional Internet access but for critical infrastructures and networks supporting services and logistical functions such as traffic sensors, lighting, security systems, and thermostats. For example, as part of its security system for the Olympics, China developed a system of video surveillance units tied together over an IPv6 network. The IPv6 "digital Olympics" network served a functional purpose, but it also sought to portray China as both an economic superpower and a global leader in advanced technologies.

### Protocols and Economic Competition

The IPv6 strategies of Asian and European Union governments shared several commonalities. First, IPv6 mandates emanated directly from national government leaders: the Japanese prime minister, Korea's minister of information and communication, India's minister of communications and information technology, the Chinese government, and the European Commission. These governments chose to mandate national upgrade strategies to promote IPv6, rather than waiting for broader markets to select IPv6 products and services. Additionally, each IPv6 promotion strategy consistently cited a twofold rationale: a recognition that each country faced a potential exhaustion of the limited resources of IPv4 addresses and an objective of becoming more economically competitive in information technology markets relative to the United States, either directly through IPv6 products, services, and expertise or through services enabled by more addresses. Additionally, governments backed national IPv6 directives with funding, tax incentives, and other direct economic inducements for service providers and equipment manufacturers. This direct governmental intervention in specific standards adoption and sweeping mandates again contravened the IETF's philosophy of working code percolating up through grassroots adoption rather than authoritative decrees. Recall that the IETF philosophy had espoused top-down mandates to be useless.[47]

IPv6 was designed to expand the number of devices able to connect to the Internet, but interestingly the objectives of national IPv6 adoption policies emphasized economic competition and nationalistic political objectives rather than the need for more IP addresses. These national strategies recognized that IPv6 adoption policies would encourage indigenous

47. David Clark et al., "Towards the Future Internet Architecture," RFC 1287, December 1991.

hardware and software manufacturers to develop new products based on IPv6 and possibly become more competitive in global IT markets as IPv6 adoption increased.

One of the reasons nations and their hardware and software vendors could seek to become competitive in offering IPv6 products is because of the implementational openness of IPv6. IPv6 is, in many ways, an example of an open standard rather than a closed, or proprietary, specification. A proprietary specification is one that is not available to manufacturers or anyone else, even for a fee. Only the company or company that developed the specification can access it and develop products based on the design specifications. The IPv6 specification, like many other Internet standards, is openly published by the IETF and available without a fee. Manufacturers have an opportunity to develop competitive IPv6 products because of the availability of the specification and because of the minimal intellectual property restrictions associated with the protocol.

The Internet has globally proliferated and provided opportunities for global economic competition and innovation, in part, because of its open protocols with minimal intellectual property restrictions such as standards-related patents. The policy of the IETF in evaluating competing technologies has traditionally been the following: "IETF working groups prefer technologies with no known intellectual property rights claims or, for technologies with claims against them, an offer of royalty-free licensing. But IETF working groups have the discretion to adopt technology with a commitment of fair and nondiscriminatory terms, or even with no licensing commitment, if they feel that this technology is superior enough to alternatives with fewer IPR [intellectual property rights] claims or free licensing to outweigh the potential cost of the licenses."[48]

Many protocols are required to implement IPv6 products, not just the IPv6 specification itself. The policy of many IPv6-related implementation strategies was to only implement protocols without restrictions on intellectual property rights (IPR). Recall that Japan's KAME Project, funded primarily by Japanese technology companies, sought to develop free IPv6 software code. The KAME Project's initial policy was to avoid any protocols with intellectual property restrictions, meaning any protocols that would require licenses to implement or that are not freely available to use. The following was KAME's first official policy on intellectual property rights:

48. Scott Bradner, editor, RFC 3979 "Intellectual Property Rights in IETF Technology," March 2005. Available at http://www.ietf.org/rfc/rfc3979.txt.

Our policy was that the KAME Project implements only protocols which:

- have no IPR [intellectual property rights] restrictions
- have IPR concerns, but are royalty-free
- do not require any license for anyone AND are free of charge for usage.[49]

The KAME Project instituted this policy to avoid intellectual property restrictions because it wanted to provide IPv6 software free of charge. If the software required royalty payments because of embedded standards-related intellectual property rights, KAME would have to charge for the software. Another concern was that, even if licensing was made available royalty-free or on so-called reasonable and nondiscriminatory terms, the project did not have a legal staff to identify and negotiate any licensing requirements for protocols.

During the development of its IPv6 software, the KAME Project software design team discovered that its software had inadvertently embedded some IPv6-related protocols with intellectual property restrictions. The design team initially removed the intellectual property restricted portions of the implementation but, after realizing the loss involved in discarding parts of the IPv6 product, decided to embark on a strategy of negotiating with patent holders to use restricted protocols without licenses.

One example of an additional IETF protocol KAME wished to use in the development of its IPv6 software was NEMO, short for Network Mobility Basic Support Protocol, a protocol extension to Mobile IPv6 (MIPv6)[50] that allows mobile networks to connect to different Internet attachment points in a manner that is transparent to the nodes connected to the mobile network.[51] Both Cisco and Nokia held intellectual property rights related to the draft NEMO specification. Cisco had pending patent applications for NEMO and specified that, if any claims of Cisco patents are necessary for implementing the standard, "any party will be able to obtain a license from Cisco to use any such patent claims under reasonable, nondiscriminatory terms, with reciprocity, to implement and fully comply with the standard."[52]

49. KAME Project statement on Intellectual Property Rights (IPR), "The KAME IPR policy and concerns of some technologies which have IPR claims." Accessed at http://www.kame.net/newsletter/20040525/.

50. See David Johnson, Charles Perkins, and Jari Arkko, "Mobility Support in IPv6," RFC 3775, June 2004.

51. Vijay Devarapalli, Ryuji Wakikawa, Alexandru Petrescu, and Pascal Thubert, "Network Mobility (NEMO) Basic Support Protocol," RFC 3963, January 2005.

52. Robert Barr, "Cisco Systems' statement about IPR claimed in draft-ietf-nemo-basic-support," June 30, 2003. Accessed at http://www.ietf.org/ietf/IPR/cisco-ipr-draft-ietf-nemo-basic-support.txt.

The issue of identifying and dealing with standards-related intellectual property rights is a significant complicating factor in any implementation of a standard, including IPv6, but the ability to openly access Internet standards and implement them with minimal intellectual property restrictions provides an opening for competitive offerings and innovation not necessarily available in sectors of information and communication technology with more restrictive approaches to standards-based intellectual property rights. Economist Rishab Ghosh suggests that a definition of open standards should address the economic effect of "supporting full competition in the market for suppliers of a technology and related products and services, even when a natural monopoly arises in the technology itself."[53]

## Cybersecurity and Distributed Warfare

While the prime minister of Japan and other government leaders touted IPv6 as part of a national economic strategy in 2000, few US institutions appeared interested in immediate IPv6 adoption. The United States already enjoyed a hegemonic information technology industry and had recently weathered the Y2K transition. The market capitalizations of Internet companies, "dot-coms," and network equipment manufacturers like Cisco and Lucent reached record valuations. Venture capital poured into companies poised to profit from web growth and Internet infrastructure expansion. The Nasdaq composite index soared more than 400 percent between 1994 and 2000. New companies such as Amazon, eBay, Google, and Yahoo! helped solidify America's dominance in Internet applications. In this context of entrepreneurship, stock market growth, and associated affluence, the prospect of the US government promoting a potentially disruptive technology upgrade seemed implausible.

US corporate Internet users had little incentive to immediately adopt IPv6 because they generally possessed ample IP addresses and an installed base of IPv4 compliant applications, network devices, and IPv4 expertise and administrative capital. Those who did face address constraints had the option of implementing network address translation (NAT), an address conservation technique that allows multiple computing devices to share a small number of Internet addresses. NAT allows a network device, such as a router, to employ limited public IP addresses to mediate between a private network with many unregistered (fabricated) IP addresses and the public Internet.

53. Rishab Ghosh, "An Economic Basis for Open Standards," December 2005. Accessed at http://flosspols.org/deliverables/FLOSSPOLS-D04-openstandards-v6.pdf.

With ample addresses and the ability to implement address conservation techniques, US businesses and the federal government were not significant IPv6 drivers relative to European and Asian policies in 2000. IPv6 advocates expressed frustration about this relative US indifference. Latif Ladid, the president and founder of an advocacy group called the IPv6 forum, criticized perceived US inaction: "As soon as IPv6 picks up in Europe, the United States will not want to miss the opportunity and will catch up. But it is an unusual situation for a country that takes leadership in practically anything; the United States seems to not be ready for it."[54]

One of the first US policy areas to even tangentially address IPv6 was Internet security. While Japan and the European Union were announcing national IPv6 strategies, one concern in the United States was the possibility of cyberterrorism, the intentional disruption or destruction of the Internet or its supporting telecommunications and power infrastructures. Increasing national dependence on information infrastructures meant that a major outage could impact critical systems like financial networks, water, power, or transportation and have significant economic and social repercussions. In 2001 several destructive Internet worms, especially Code Red and Nimbda, resulted in disruptive and costly Internet outages.

Within the context of increasingly virulent computer worms and economic and social dependence on networks, the September 11, 2001, terrorist attacks on the United States crystallized an already mounting concern about the vulnerability of economically and operationally vital information networks to possible cyberterrorism. One governmental response to this concern was the development of the *National Strategy to Secure Cyberspace*, the culmination of a lengthy analysis seeking a reduction in US vulnerability to attacks on critical information infrastructures. One of the Strategy's recommendations included improving the security of several network protocols,[55] including the Internet Protocol. The strategy noted that Japan, the European Union, and China were already upgrading from IPv4 to IPv6 and cited "improved security features,"[56] as one of the benefits of IPv6, although Richard Clarke, the top counterterrorism official at the

54. Quoted in "The Numbers Game" by Reed Hellman on the IPv6 Forum website. Accessed on www.ipv6forum.org on October 1, 2002.

55. The network protocols addressed in the "National Strategy to Secure Cyberspace" (February, 2003) included the domain name system (DNS), border gateway protocol (BGP), and the Internet protocol (IP), p. 30. Accessed at http://www.whitehouse.gov/pcipb/cyberspace_strategy.pdf.

56. "National Strategy to Secure Cyberspace," February 2003, p. 30.

time of the September 11 attack and later the "cybersecurity czar," noted that "a world of mixed IPv4 and IPv6 implementations actually increases the security threat."[57] IPv6 received only a cursory mention in the strategy, but the document asserted as a fact that IPv6 was more secure than IPv4. One of the document's concrete recommendations called for the US Department of Commerce to launch a task force examining issues related to IPv6.[58]

What seemed like a significant momentum shift also occurred on June 9, 2003, when the US Department of Defense mandated it would transition to IPv6 by 2008. John Stenbit, then assistant Secretary of Defense for Networks and Information Integration and DoD chief information officer, issued a memorandum establishing the directive, which stated, "The achievement of net-centric operations and warfare, envisioned as the Global Information Grid (GIG) of inter-networked sensors, platforms, and other Information Technology/National Security System (IT/NSS) capabilities (ref a), depends on effective implementation of IPv6. . . . "[59]

The DoD's rationale for upgrading to IPv6 was multifaceted. On one hand, the formal memorandum announcing the IPv6 mandate cited the requirement for end-to-end security and management and more addresses for military combat applications.[60] On the other hand, Stenbit's press briefing[61] described how IPv4 had three major shortcomings: end-to-end security, quality of service, and address shortages. Only two of these were important to the DoD. The one he described as not salient to the DoD was IP address shortages, although Stenbit acknowledged this was important

57. Reported in *Converge Network Digest*. Accessed at http://www.convergedigest .com/packetsystems.html, 2002 on November 13, 2002.

58. *National Strategy to Secure Cyberspace* recommendation A/R 2-3: "The Department of Commerce will form a task force to examine the issues related to IPv6, including the appropriate role of government, international interoperability, security in transition, and costs and benefits. The task force will solicit input from potentially impacted industry segments." February 2003, p. 56.

59. United States Department of Defense Memorandum issued by DoD chief information officer, John P. Stenbit for Secretaries of the Military Departments, Subject: Internet Protocol Version 6 (IPv6), June 9, 2003. "ref a" refers to DoD 8100.1 Global Information Grid Overarching Policy, September 19, 2002. Accessed at http://www .dod.gov/news/Jun2003/d20030609nii.pdf on July 20, 2003.

60. Ibid.

61. See "Briefing on New Defense Department Internet Protocol," John Stenbit, Presenter, Friday, June 13, 2003. Accessed at http://www.dod.mil/transcripts/2003/ tr20030613-0274 on July 20, 2003.

to Europe. The shortcomings of concern to the DoD were end-to-end security and quality of service. Consistent with the US *Strategy to Secure Cyberspace* and the promise of IPv6 in the EU and some Asian countries, the DoD IPv6 strategy cited enhanced security as one rationale for transitioning to IPv6. Defense Department discussions about IPv6 emphasized its ability to keep military personal safe and secure in a new, fluid, and distributed battleground.

The new DoD policy specified that, beginning in October 2003, all information technology products procured or developed must be IPv6 capable. One open issue was the definition of IPv6 capable. In 2003 many software and hardware products contained native IPv6 capability as well as IPv4. Purchasing these products did not equate to implementing IPv6. The term IPv6 capable seemed malleable, ranging from procuring routers and operating systems already including dormant IPv6 support, versus implementing IPv6 as the network-layer protocol along with IPv4 through complicated dual stack IPv6 and IPv4 software implementations or protocol tunneling.

The DoD IPv6 decision, like the publication of the *Strategy to Secure Cyberspace*, occurred contextually in the aftermath of the September 11, 2001, terrorist attacks on the United States. The IPv6 decision appeared interleaved with a broader conversation about the war on terrorism, framed as a new type of war requiring distributed rather than centralized information flows, mobile versus static command and control, and a ubiquitous versus defined front. The new type of war required a new strategy, the Global Information Grid (GIG), which required a new standard, IPv6. The DoD incorporated the GIG/IPv6 strategy within its Joint Transformation Roadmap designed to transform the military into a force geared toward supporting the DoD's top priorities. These priorities included improving intelligence gathering, surveillance, and strike capabilities in fighting the global war on terrorism, and empowering "warfighters in the distributed battlespace of the future."[62]

The promise of IPv6 appeared to fit in with the political objectives for a distributed, decentralized, vision of fighting a ubiquitous war on terrorism. Cold war network infrastructure approaches had focused on centralized command and control,[63] but the new GIG architecture emphasized distributed and ubiquitous sensors and decision making.

62. United States Department of Defense Joint Transformation Roadmap, January 21, 2004. Accessed at http://www.ndu.edu/library/docs/jt-transf-roadmap2004.pdf on April 4, 2004.
63. Paul Edwards, *The Closed World: Computers and the Politics of Discourse in Cold War America*, Cambridge: MIT Press, 1996.

## Concerns about US Economic Competitiveness

Despite the DoD's IPv6 commitment, overall US government views about the extent of federal IPv6 involvement varied by agency. For example, the Commerce Department's stance on IPv6 seemed cautious relative to the DoD's position. One of the directives in President Bush's *National Strategy to Secure Cyberspace* had called for a formal examination of IPv6 issues. The Commerce Department convened a task force assessing the appropriate role of the US government in IPv6 deployment and evaluating possible economic opportunities. The National Institute of Standards and Technology (NIST) and the National Telecommunications and Information Administration (NTIA) co-chaired the task force and solicited public input about US IPv6 opportunities, the state of international and domestic IPv6 deployments, technical and economic IPv6 issues, and the merits of US federal government involvement in IPv6.[64] The Commerce Department task force received twenty-one public responses, many from American software, hardware, and IT services vendors, including Bell South, Sprint Corporation, Microsoft Corporation, Qwest Communications, VeriSign, WorldCom, and Motorola. The task force also received public responses from a few individuals in the Internet standards and IP address registry communities and several advocacy institutions, including the Electronic Privacy Information Center (EPIC), the North American IPv6 Task Force (NAv6TF), and the Internet Security Alliance (ISA).

The Commerce Department's task force published a draft discussion report, "Technical and Economic Assessment of Internet Protocol Version, 6 (IPv6),"[65] generally concluding that market mechanisms, not the federal government, should drive IPv6 adoption. The task force acknowledged that most major software and hardware products, like the Linux operating system, some Microsoft products, and Cisco and Juniper routers, already embedded IPv6 capability, but that these features were generally dormant and not activated by users. NTT/Verio was the only service provider already

64. United States Department of Commerce, "Commerce Department Task Force Requests Comments on Benefits and Costs of Transition to New Internet Protocol, Appropriate Role of Government in IPv6 Deployment to Be Addressed," Washington, DC, January 15, 2004. Accessed at http://www.ntia.doc.gov/ntiahome/press/2004/IPv6_01152004.htm.

65. United States Department of Commerce, "Technical and Economic Assessment of Internet Protocol Version 6 (IPv6)," Washington, DC, July, 2004. Accessed at http://www.ntia.doc.gov/ntiahome/ntiageneral/ipv6/final/ipv6final.pdf on September 17, 2004.

offering IPv6-based Internet access service. The United States had an enormous installed base of IPv4-based communications, and the Commerce Department report estimated that less than 1 percent of US Internet users employed IPv6 services.

Considering the enormous installed base of IPv4 and the transition costs for upgrading from IPv4 to IPv6, a major policy question was whether the benefits of IPv6 outweighed the expense of an accelerated, government-influenced or government-funded conversion to IPv6. ISPs would incur the highest transition costs, related to upgrading hardware and software and the cost of acquiring IPv6 expertise, while envisioning scant demand in the United States and therefore no return on investment. The Commerce Department analysis concluded that many of the touted benefits of IPv6 were already available in IPv4: "IPv4 can now support, to varying degrees, many of the capabilities available in IPv6."[66]

For example, IPv6 advocates touted improved security as a benefit because the IPv6 standard called for the support of an encryption protocol, IPsec. In contrast, the Commerce Department task force noted that, while "IPsec support is mandatory in IPv6. IPsec *use* is not"[67] and that IPv4 networks can also use IPsec encryption. IPv6 might actually be less secure than IPv4. The analysis summarized the security issue as follows:

[It] is likely that in the short term (i.e., the next 3 to 5 years) the user community will at best see no better security than what can be realized in IPv4-only networks today. During this period, more security holes will probably be found in IPv6 than IPv4.[68]

In addition to dismissing improved security as an incentive for upgrading, the report also concluded that many existing mechanisms already mitigated address depletion problems.

Another concern was whether the United States would somehow become disadvantaged economically because of more rapid IPv6 dissemination internationally through governmental promotion and incentives in Asia and Europe. On one hand, the Commerce Department argued that major US software and hardware vendors already supported both IPv4 and IPv6 and sold IPv6 products in international markets. Lethargic US IPv6

66. United States Department of Commerce, "Technical and Economic Assessment of Internet Protocol Version 6 (IPv6)," Washington, DC, July, 2004, Introduction, p. 2. Accessed at http://www.ntia.doc.gov/ntiahome/ntiageneral/ipv6/final/ipv6final .pdf.
67. Ibid., ch. 2, p. 6.
68. Ibid.

adoption would not alter the opportunity for American technology companies to compete in these global markets. Conversely, concerns about the shift of intellectual resources to Asia in well-funded IPv6 research and development fit into broader Commerce Department and social concerns about the outsourcing of IT jobs to India, China, and other nations. Despite overall outsourcing concerns, the Commerce Department's draft report concluded that, while the US government could "stimulate adoption" as an IPv6 customer, ultimately private sector decisions should drive the market.

The Commerce department's laissez-faire conclusions faced ardent criticism from US IPv6 advocates, who questioned the prospects of future US economic competitiveness in light of rapid international IPv6 deployment. IPv6 advocates criticized the Commerce Department recommendation to allow markets to determine IPv6 deployment and questioned where the United States would be economically without a history of information technology investment in such areas as telegraph lines, digital computers, satellites, radar, and early Internet innovations such as packet switching and the original ARPANET research project. Alex Lightman, a prominent IPv6 advocate and chairman of IPv6 Summits, Inc., suggested that IPv6 investment might stave off unemployment and might generate 10 million new jobs.[69] Achieving this, he argued, would require $10 billion in government investment over four years and a federal mandate that all its systems transition to IPv6. This type of a mandate would be more contained than national policies in China, Korea, and Japan mandating that all systems, not just federal IT systems, deploy IPv6.

What was at stake if the United States failed to upgrade to IPv6 while other parts of the world, especially China, India, Korea, Japan, and the European Union upgraded to IPv6? Lightman argued that US exports of Internet products were at risk to such an extent that the United States would one day retrospectively ask "Who lost the Internet?"[70] The Commerce Department report noted that US software and hardware vendors generally supported both IPv4 and IPv6, primarily because they served global markets, not just US markets. Yet IPv6 advocates seemed to be suggesting that the IPv6 issue have a Sputnik-like urgency for the federal government.

69. Alex Lightman, "10 Million New Jobs from IPv6: The Case for U.S. Government Investment," *6Sense Newsletter*, November 2004.
70. Alex Lightman, "Lead, Follow, or Lose the Great Game: Why We Must Choose a U.S. IPv6 Leader," *6Sense Newsletter*, April 2005.

## Protocol Hearing on Capitol Hill

Concerns about IPv6 and American IT competitiveness and outsourcing threats escalated to the US Congress in June 2005, exactly five years after Japanese Prime Minister Yoshiro Mori announced his country's e-Japan program establishing the goal of a nationwide IPv6 upgrade. Virginia Representative Tom Davis (R), chairman of the Government Reform Committee, convened a congressional committee hearing on the Internet and IPv6. The hearing, "To Lead or Follow: the Next Generation Internet and the Transition to IPv6," examined questions about economic opportunities and risks to the United States and about the possibility of a mandate to upgrade the federal government to IPv6.

Representative Davis opened the congressional hearing with remarks about the relationship between the geographical area he represented and the Internet. Davis asserted that 25 percent of the world's Internet service providers were within an hour's drive of Fairfax County, Virginia and that 25 percent of Internet traffic passed through a hub in northern Virginia. The Representative further stated that "the current Internet, and the protocols and networks that underpin it, may have reached its limits."[71] The hearing generally assumed that the Internet required upgrading and Davis wished to understand the economic implications of Asia's lead, particularly China's lead, in investing hundreds of millions of dollars in aggressive IPv6 deployment. In addition to concerns about US Internet competitiveness, Davis mentioned homeland security and US defense capability as possible drivers for examining IPv6. Seven individuals offered testimony in the IPv6 hearing, but notably missing were any individuals speaking on behalf of US Internet users, whether corporate, institutional, or individual. Also missing were individuals involved in standards development, with the exception of John Curran testifying for Internet registrar ARIN, but who had served on the IPng Directorate responsible for selecting IPv6 from competing alternatives.

The prospect of the United States trailing Asia in Internet innovation, jobs, and economic stature thematically dominated the hearing. Lightman's testimony contained the most emphatic caveats about the economic and political stakes of IPv6. According to Lightman, federal leadership in IPv6, particularly a mandate to transition federal systems to IPv6, might

71. From the opening statement of Chairman Tom Davis, "To Lead or to Follow: The Next Generation Internet and the Transition to IPv6," Committee on Government Reform, Washington, DC, June 29, 2005.

create 10 million American jobs, generate trillions of dollars in revenue, and add products vital to national defense, homeland security, and network security.[72] Conversely, government inaction would result in lost jobs and market share. He also underscored the imbalance between US and international IPv6 expenditures, suggesting that China, Japan, Korea, and the European Union had invested $800 million versus the US committing $8 million.

The absence of corporate, institutional, or individual Internet users in the congressional hearings accentuated the disconnect between advocacy about upgrading to IPv6 in the United States and the reality of what the professionals responsible for network protocol upgrades were actually doing. For example, a 2005 survey of government and private sector information technology managers about IPv6 plans revealed two circumstances: (1) among both private and public technical personnel, there were "low levels of interest in IPv6," and (2) despite the DoD IPv6 mandate, federal government information technology professionals demonstrated a lower level of IPv6 awareness than even disinterested corporate professionals. The survey further underscored a lack of consensus about the meaning of "IPv6-ready," ranging from IPv6 software in all applications, network devices, and infrastructural components comparable to IPv4 features, to the belief, expressed by 37 percent of respondents, that IPv6-ready meant the product should be upgradeable to IPv6 at some future time. The surveyed information technology professionals overwhelmingly doubted IPv6 would help them achieve their organizations' IT objectives and failed to see a compelling functional or budgetary reason to upgrade. Those that did see a compelling reason cited what they perceived as improved security of IPv6.[73]

But by August 2005 the Office of e-Government in the Office of Management and Budget issued a memorandum directing that agencies should upgrade their agency backbones to be IPv6 capable by June 30, 2008.[74] An IPv6-compliant system, in this case was defined as able to "receive, process,

72. Alex Lightman, Testimony submitted to the Committee on Government Reform Hearing, "To Lead or Follow: The Next Generation Internet and the Transition to IPv6," Washington DC, June 28, 2005.

73. Source: Juniper Networks 2005 Federal IPv6 IQ Study. Accessed at http://209.183.221.252/Juniper_Networks_2005_Federal_IPv6_IQStudy.pdf on December 19, 2005.

74. See the "Memorandum for the Chief Information Officers," M-05-22 from Karen S. Evans, the Administrator of the Office of E-Government and Information

and transmit or forward (as appropriate) IPv6 packets and should interoperate with other systems and protocols in both IPv4 and IPv6 modes of operation."[75]

## Protocols as Social Intervention

In addition to strategies focused on economic and national competitiveness rationales for IPv6, many IPv6 advocates have also situated the protocol in a more explicitly moral space, linking the protocol with promises of democratization, freedom, social justice, and third world development. Other technology standards have been similarly linked to social objectives. Ken Alder describes how, two hundred years earlier, French Revolutionary scientists viewed the metric standard as a utopian democratic vision of equal access to information versus powerful entities wishing to protect their interests. Expectations about the social benefit of the expansion of the Internet address space under IPv6 have also mirrored descriptions of the expansion of "ether" (electromagnetic spectrum) in radio broadcasting a century earlier. In *Inventing American Broadcasting*, Susan J. Douglas discusses the "democratic rhetoric that described the air as being free and the property of the people."[76] Hugh R. Slotten, in *Radio and Television Regulation in Broadcast Technology in the United States, 1920–1960*, explores the utopian rhetoric surrounding technological advancements in radio broadcasting. Engineers and policy makers, as well as some public participants, viewed broadcasting innovations as precursors to social progress and as imperatives for solving social problems.[77]

Claims about IPv6 as a solution to social problems followed a similar trajectory. The following (abridged) posting appeared on the opening web page of the North American IPv6 Task Force (NAv6TF): "IPv6 is about Freedom. I agree. . . . Today, the cost of freedom is great. IPv6 reduces that cost I believe greatly, thus IPv6 is also about peace. And peace is good for

Technology, subject: "Transition Planning for Internet Protocol Version 6 (IPv6), August 2, 2005. Accessed at http://www.whitehouse.gov/omb/memoranda/fy2005/m05-22.pdf.

75. Ibid.

76. Susan J. Douglas *Inventing American Broadcasting: 1899–1922*. Baltimore: Johns Hopkins University Press, 1987, p. 214.

77. Hugh R. Slotten, *Radio and Television Regulation: Broadcast Technology in the United States, 1920–1960*. Baltimore: Johns Hopkins University Press, 2000, p. 237.

business. So from a business perspective the cost of not doing IPv6 is great. This should be part of our business view."[78]

The NAv6TF's mission and IPv6 vision reflected the objectives of its parent organization, the IPv6 Forum. Latif Ladid founded the IPv6 Forum in May 1999, shortly after the formal ratification of the IPv6 specifications, to promote worldwide deployment of IPv6. In presentations about IPv6, Ladid has often suggested that participants promote IPv6 to generally serve society. He has argued that IPv6 could help alleviate the digital divide and suggested that those interested in IPv6 "do something for yourself, your community, your society, your country, your world. Be a pioneer in IPv6."[79]

IPv6 advocates have worked directly with governmental agencies around the world, including some US entities including the US DoD and members of Congress. From its 2001 inception as a North American outgrowth of the IPv6 Forum, the NAv6TF worked with US government entities to promote IPv6, assess possible roles of IPv6 in the federal government, and address technology deployment issues. As part of this liaison the institution participated in "Moonv6," a collaborative IPv6 test pilot launched in 2003 with the InterOperability Laboratory at the University of New Hampshire, the US Department of Defense Joint Interoperability Testing Command, and industry vendors. The founding mission of the collaboration sought to develop a test bed network demonstrating interoperability between diverse IPv6 products. Moonv6 project leaders reflected a mixture of IPv6 perspectives and included: NAv6TF Chair and IETF contributor Jim Bound; Major Roswell Dixon, IPv6 Action Officer within the DoD's Joint Interoperability Test Command; and Yasuyuki Matsuoka of NTT in Tokyo, Japan. The test bed's nomenclature "Moonv6" symbolically represented the importance participants placed on IPv6. In a meeting discussing the seriousness with which the United States should consider IPv6, someone questioned whether the United States should view IPv6 with the same urgency it viewed reaching the moon in 1969.[80] The IPv6 test bed leaders selected the name "Moonv6" accordingly.

78. Jim Bound, a Hewlett Packard fellow who served as the chair of the IPv6 Forum Technical Directorate, chair of the North American IPv6 Task Force, and who had previously served within the IETF on the IPng Directorate, posting on North American IPv6 Task Force website. Accessed at www.nav6tf.org on October 15, 2002.
79. Latif Ladid speaking at the U.S. IPv6 Summit, Arlington, VA, December 10, 2003.
80. According to the Moonv6 website. Accessed at www.moonv6.org on October 15, 2004.

A variety of optimistic expectations for IPv6 similarly converged at a one day public IPv6 meeting in July 2004, entitled "Deploying IPv6: Exploring the Issues." The US Commerce Department sponsored the meeting, which included Vinton Cerf, Mark Rotenberg of the Electronic Privacy Information Center (EPIC), various representatives from industry, academe, and government, and IPv6 advocates Latif Ladid and Jim Bound. Jim Bound posed the following provocative question to the morning session panelists: How can IPv6 help "the social aspects that we face in our own inner city ghettos, for security defense networks[?] In 9/11, police, port authority, and firemen were unable to communicate. That cost lives. That's a social problem, too. And how can IPv6 maybe help it so that the kids that I work with in my private life from the inner city ghettos have equal opportunity to learn about communications, learn about the Internet, and evolve?"[81] Bound's question suggested an association between IPv6 and a broad range of social concerns: poverty, national defense, homeland security, first-responder capability, and education.

Not everyone embraced expectations about the broad social benefits of IPv6. Paul Francis of Cornell University characterized the linkage between social inequity, ghettos, and IPv6 as tenuous and Mark Rotenberg of EPIC summarized: "It's a bit of a stretch to think that we solve problems of social inequality through IPv6 deployment."[82] In contrast, Bound's colleague, Latif Ladid, accentuated the social possibilities of IPv6 and portrayed implementing the standard as a moral obligation: "I think we have a moral obligation and a unique opportunity to do something special, not only to look at the profits and look at the stock market, and so on and so forth. I think we've got to go beyond this and do something that's going to give some kind of hope and vision for the entire world . . . most probably the kids in Detroit and the Bronx, so on and so forth, they have exactly the same digital chasm that we have in Africa."[83]

Ladid's choice of the term "moral obligation" toward the next generation of children and Bound's references to inner city ghettos certainly appear distant objectives from the DoD's distributed warfare strategy or the

81. Transcript of the Department of Commerce public meeting, "Deploying IPv6: Exploring the Issues," Washington, DC, July 28, 2004. Accessed at http://www.ntia .doc.gov/ntiahome/ntiageneral/ipv6/IPv6Transcript_part1.htm.

82. From the written transcript of the Department of Commerce public meeting, "Deploying IPv6: Exploring the Issues," Washington, DC, July 28, 2004.

83. Latif Ladid statement from the transcript of the Department of Commerce public meeting, "Deploying IPv6: Exploring the Issues," Washington, DC, July 28, 2004.

economic objectives of Japan and the European Union. Nevertheless, themes of IPv6 improving children's lives and ameliorating social problems accompanied various IPv6 rationales. Even the director of Architecture and Interoperability for the US DoD, in public remarks, had suggested that IPv6 "is really important to the lives of kids."[84] His statement mirrored the IPv6 advocacy rhetoric of Bound and Ladid in indicating that IPv6 would improve children's lives. These rationales alluded to IPv6 as a moral intervention.

### Questioning IPv6 Security[85]

One common thread within IPv6 advocacy was the espousal of "increased security" as a considerable advantage of IPv6 over IPv4. The 2003 US Defense Department memorandum mandating IPv6 cited end-to-end security as one rationale for upgrading. The US *Strategy to Secure Cyberspace* had described IPv6 as providing greater security than IPv4. Japan's IT Strategy Council argued that a benefit of IPv6 was its enhanced security features. The IPv6 Forums claimed that security features, as well as address space expansion, was sufficient justification for upgrading. IPv6 advocates have consistently reproduced this argument and the technical media has unquestioningly depicted the security benefits of IPv6. For example, the networking industry journal, *Network World*, argued: "IPv6 promises a dramatically larger addressing scheme as well as enhanced security and easier administration."[86] Technical engineers for vendors economically invested in IPv6 have also touted security as an inherent IPv6 feature.

Despite these claims, IPv6 does not appear to inherently provide greater security. Rather than providing "improved security" or specifically addressing security at all, IPv6, a less mature protocol than IPv4, actually raises some security issues. As with most developing protocols, security weaknesses have been identified in IPv6-enabled products, and these require user action to mitigate. IPv6 capability is present in many products, even

84. John L. Osterholz, director of Architecture and Interoperability, US Department of Defense, keynote address at US IPv6 Summit, Arlington, VA, Tuesday, December 9, 2003.
85. See Laura DeNardis, "Questioning IPv6 Security," 36 *Business Communications Review* 51–53 (2006).
86. Carolyn Duffy Marsan, "IPv6 Expert Sees Adoption Growing . . . Slowly," *Network World*, September 27, 2004.

if left dormant by users. Those not specifically deploying IPv6 capabilities might assume the security vulnerabilities and associated patches are not pertinent to their network environments and forgo the necessary network security responses.

Some groups within the US government have questioned the extent to which IPv6 provides greater security than IPv4. A 2005 Government Accountability Office (GAO) analysis of IPv6 identified security risks as a significant transition consideration for federal agencies. The US House of Representatives Committee on Government Reform requested that the GAO perform an analysis auditing the progress the DoD and any other government agencies have made in transitioning to IPv6 and identifying considerations for agencies upgrading or planning to upgrade. The GAO methodology employed government auditing standards and issued its findings in a May 2005 report, entitled "Internet Protocol Version 6: Federal Agencies Need to Plan for Transition and Manage Security Risks."[87]

The GAO noted the dormant IPv6 capability in the software and hardware products many federal agencies already routinely procured. Most routers already incorporated features, by 2005, allowing users to configure networks for IPv6 traffic. Similarly, leading operating systems such as Linux, Solaris, Cisco IOS, Microsoft Windows, and Apple OS X supported IPv6. The GAO report stressed that this dormant IPv6 capability actually exacerbated rather than mitigated security risks. For example, an employee enabling IPv6 capability might create an inadvertent security problem because an institution's security system configuration might not detect breaches exploiting IPv6 features. The GAO audit specifically investigated two IPv6 characteristics, automatic configuration and tunneling, for security vulnerabilities. The audit confirmed already widely understood security vulnerabilities of these features and determined "they could present serious risks to federal agencies."[88] Protocol designers included automatic configuration as an IPv6 feature intended to simplify network administration of IP addresses. This autoconfiguration feature might permit an unauthorized router connected to an agency network to reconfigure neighboring system addresses and routers, exposing them to vulnerabilities because

87. US Government Accountability Office, Report to Congressional Requesters, "Internet Protocol Version 6, Federal Agencies Need to Plan for Transition and Manage Security Risks," GAO-05-471, May 2005.
88. Ibid., p. 22.

resulting IPv6 activity could circumvent existing intrusion detection systems (IDS). The GAO audit similarly assessed security vulnerabilities associated with tunneling, the technique of transmitting IPv6 packets over an IPv4 network. The embedding of IPv6 formatted information within IPv4 packets allowed potentially unauthorized activity to occur undetected by firewalls.

The US Computer Emergency Readiness Team (US-CERT) also identified numerous IPv6 security vulnerabilities. CERT, originally an acronym for Computer Emergency Response Team, formed in the aftermath of the 1988 computer worm that disrupted thousands of Internet-connected computers. The worm, launched by Cornell graduate student Robert Morris, raised awareness about network security vulnerabilities and led to DARPA establishing a new DoD-funded organization at Carnegie Mellon University called the Computer Emergency Response Team to respond to security incidents and educate users.[89] Years later, in September 2003, the US Department of Homeland Security created a new CERT, the US-CERT, which would supersede but coordinate with the Carnegie Mellon operated CERT and numerous other CERT organizations throughout the world. The formation of US-CERT reflected homeland security concerns about cyberterrorism in the wake of the September 11 attacks and awareness of increasing economic and political value of the Internet as a critical national infrastructure. As part of its activities, US-CERT identified vulnerabilities in products, systems, and protocols and identified a number of inherent security vulnerabilities in the IPv6 protocol.

To provide a few selected examples, the following are some abridged CERT vulnerability notes addressing a historical range of IPv6-related security weaknesses:

Cisco IOS IPv6 denial-of-service vulnerability
(Vulnerability note VU472582)[90]

- A vulnerability in the way Cisco IOS handles IPv6 packets could result in a remotely exploitable denial of service.
- A remote attacker may be able to cause an affected device to reload, thereby creating a denial of service condition.

89. DARPA press release, "DARPA Establishes Computer Emergency Response Team," December 6, 1988.
90. US-CERT Vulnerability note VU472582, "Cisco IOS IPv6 Denial-of-Service Vulnerability," *First Public*, January 26, 2005.

Juniper JUNOS Packet Forwarding Engine (PFE) IPv6 memory leak
(Vulnerability note VU658859)[91]

▪ The Juniper JUNOS Packet Forwarding Engine (PFE) leaks memory when certain IPv6 packets are submitted for processing.
▪ If an attacker submits multiple packets to a vulnerable router running IPv6-enabled PFE, the router can be repeatedly rebooted, essentially creating a denial of service for the router.

Solaris Systems May Crash in Response to Certain IPv6 Packets
(Vulnerability note VU658859)[92]

▪ Solaris 8 systems that accept IPv6 traffic may be subject to denial of service attacks from arbitrary remote attackers.

IPv6 is a less mature protocol than IPv4, so the ongoing identification of protocol-specific product vulnerabilities is not unusual. Each vulnerability pronouncement necessitates that users install vendor issued software patches and upgrades.

In some cases users were not even cognizant of the IPv6 features inherent in products, a phenomenon the GAO's IPv6 assessment emphasized. Many users assumed IPv6 security advisories were not applicable unless they had activated IPv6 features and would assume vulnerability announcements did not pertain to their systems.

Even if there were no protocol vulnerabilities within IPv6, it is important to note that the protocol does not intrinsically address security issues. One of the reasons for the linkage between IPv6 and improved security is the historical association between IPv6 and a separate network-layer protocol, IPsec. The early Internet and its predecessor networks involved relatively closed information exchange among trusted individuals. As the Internet began to expand and after network security vulnerabilities and disruptions began to occur, security became more of a concern to Internet technical designers. During the development of IPv6 in 1990, security was a significant design consideration and Internet designers decided to mandate the use of IPsec in the draft IPv6 specifications. This connection between early drafts of the IPv6 protocol and IPsec is one origin for ongoing claims that IPv6 provides enhanced Internet security over IPv4.

But there are several circumstances suggesting that IPv6 is not inherently more secure. First, IPSec encryption can easily be implemented in IPv4

91. US-CERT Vulnerability note VU658859, "Juniper JUNOS Packet Forwarding Engine (PFE) IPv6 Memory Leak," *First Public*, June 29, 2004.
92. US-CERT Vulnerability note VU370060, "Solaris Systems May Crash in Response to Certain IPv6 Packets," *First Public*, July 21, 2003.

networks as well as in IPv6 networks. The argument that IPSec improves the security of IPv6 networks is equivalent to the argument that IPSec improves the security of IPv4 networks. It is the encryption provided by IPSec that provides security, not the IPv4 protocol or the IPv6 protocol. Second, the later (1998) IPv6 specification was updated to eliminate the mandatory inclusion of IPsec with IPv6. Third, just because a security technique is mentioned or recommended in a protocol specification does not mean that it will automatically be included in a product implementation of that protocol. Finally, even when IPsec is implemented within IPv6 networks, this is only one aspect (encryption) of a broader security framework required to protect against worms, viruses, distributed denial of service attacks, and other security threats.

Claims about IPv6 improving network security also usually assume that a network implementation will exclusively deploy end-to-end IPv6 and eliminate IPv4. If IPv4 were eliminated, this would obviate the need for network address translation, the IP address conservation technique that allows numerous devices to share IP addresses. There have historically been security concerns about the deployment of NAT devices because they represent an information intermediary that interrupts the end-to-end architecture of the Internet that locates intelligence in network end points. (Others argue that NAT can sometimes improve security by obscuring a private network's internal Internet addresses.) Regardless, the promise of end-to-end IPv6, and the associated obviation of NAT devices, is unlikely because IPv4 and IPv6 will likely coexist in most networks, with the exception of relative closed network environments that can exclusively use IPv6.

Furthermore, the approaches for transitioning to IPv6, described in detail later, each present a different set of complexities and security considerations. Rather than simplifying security, mixed protocol environments can actually complicate security. IPv6, like most evolving protocols, has experienced its share of intrinsic security vulnerabilities. But claims that IPv6 improves security are misleading.

### Reality Check on IPv6 Deployments

Considering the history of optimistic IPv6 expectations and aggressive adoption plans, how have strategic plans progressed? Japan's IT Strategy ranked among the most aggressive for implementing IPv6. Recall that Japanese Prime Minister Yoshiro Mori established a 2005 deadline for upgrading every Japanese business and public sector computing device to IPv6. The e-Japan program sought to elevate Japan to a global IT leader by 2005, an objective requiring a complete national transition to IPv6.[93] By 2005 this transition had simply not occurred. According to the official description from Japan's IPv6 Promotion Council in 2005, "The spread of IPv6 has just begun" and "there are still a number of barriers to the deployment of IPv6 and promotion measures to solve this problem and remove the barriers are needed for some time. As we pull through this stage, IPv6 will propagate on its own."[94]

For the introduction period of IPv6 the Council noted that they could not expect to achieve "things only IPv6 can do,"[95] acknowledging that IPv6 is not an application but a transparent network addressing and routing protocol. It also noted that IPv4 and IPv6 would coexist and that IPv6 security issues were complex. Korea's IPv6 deployment status in 2005 also primarily involved trial networks. In 2005 Korea's IPv6 strategy was modified to continue research and development test networks and expand commercial services toward a goal of full national IPv6 deployment by 2010.[96] European Union, Chinese, and Indian IPv6 deployments were similarly inchoate. The overall worldwide status of IPv6 deployment, while progressing slightly, still primarily involved measured network pilots. Limited production networks were beginning to become available but, as IPv6 advocate Jim Bound described, not with "the required management, application, middleware, or security infrastructure required for most production networks."[97] In the US government, backbone networks became "IPv6 compliant," but this has not necessarily translated into IPv6 use. Some of the Internet's DNS, in 2008, was upgraded to support IPv6. In 2008, ICANN added IPv6 capability for six of the thirteen root servers, allowing for improved IPv6 usage of the DNS.[98]

93. Specified in the e-Japan Priority Policy program, Policy 2, March 20, 2001. Accessed at http://www.kantei.go.jp/foreign/it/network/priority/slike4.html on April 15, 2003.

94. IPv6 Promotion Council of Japan, "2005 Version IPv6 Deployment Guideline: About the IPv6 Deployment Guideline," March 2005. Accessed at http://www.v6pc .jp/pdf/en-01-IPv6_Deployment_Guideline.pdf on December 4, 2005.

95. Ibid.

96. "IPv6 Development Status in Korea," Doc no: Telwg31/IPv6/05; APEC telecommunications and Information Working Group 31st Meeting, Bangkok, Thailand, April 2005.

97. Jim Bound, "IPv6 Deployment State 2005," in 6 *Upgrade: The European Journal for Informatics Professionals* (April 2005).

98. See ICANN announcement "IPv6 Address Added for Root Servers in the Root Zone: Addition Enhances End-to-End Connectivity for IPv6 Networks," February 2008. Accessed at http://www.icann.org/en/announcements/announcement-04feb08.htm.

The historical narrative about Internet addresses has been that IPv4 addresses are quickly depleting and that the upgrade to IPv6 is imminent. The historical reality has been that the pool of Internet addresses has, in fact, been nearly depleted but this phenomenon has not been accompanied by any significant upgrade to IPv6. There is nothing surprising about the historical trajectory of increasing Internet address assignments as more people and devices become connected to the Internet and as new mobile applications and computing devices require more and more Internet addresses.

What has been considered surprising is the sluggishness of IPv6 deployment and adoption, especially considering the national IPv6 mandates from governments. In 2008, after many IPv6 implementation deadlines had come and gone, Internet engineer Geoff Huston estimated that, based on web server access data: "the relative rate of IPv6 use appears to be around 0.3 percent of the IPv4 use, or a relative rate of 3 per 1,000."[99]

Many (but not all) popular applications, operating systems, and hardware devices have IPv6 capability, but this has not yet translated into extensive IPv6 implementations or use. It is difficult to avoid drawing analogies between IPv6 and the history of OSI protocols, which were embraced in national strategies and celebrated internationally but which never flourished.

## Protocol Transition Challenges

Part of the difficulty is the reality that an IPv6-only Internet device *cannot* reach an IPv4-only device directly (see figure 4.1). In other words, a laptop connected to the Internet via IPv6-only cannot directly reach an IPv4 web server (e.g., popular news or social networking sites). To illustrate the end result to Internet users, if a website such as cnn.com does not include IPv6 support, those accessing the web from IPv6 networks cannot reach these websites. Yet IPv4-only sites are the norm and will likely remain the norm for the foreseeable future. Most IPv6 deployments must include a technique for reaching these IPv4-only sites. It may be the case that some potentially "walled garden" applications such as VoIP, interactive gaming, and Internet Protocol Television (IPTV) applications offered by service providers will not require backward compatibility with IPv4 because these

99. Geoff Huston, "IPv6 Deployment: Just Where Are We?" Featured posting on the CircleID website, March 31, 2008. Accessed at http://www.circleid.com/posts/ipv6_deployment_where_are_we/.

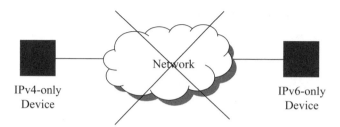

**Figure 4.1**
IPv4-only device cannot directly reach IPv6-only device

may be self-contained systems that can use end-to-end IPv6 with no connectivity to legacy IPv4 protocols. But many IPv6 deployments require the ability to communicate with servers and devices already connected to the Internet via IPv4.

IPv6 is not backward compatible with IPv4. The IPv6 header, the control and addressing fields accompanying information as it traverses a network in a packet, has its own distinct set of fields and formatting structures. If devices using the IPv6 protocol receive information using IPv4 formatting, these devices cannot natively process this information without some form of technical translation, including translating the different sized (32-bit and 128-bit) source and address destination addresses between IPv4 and IPv6.

An IPv6-only device communicating with an IPv4-only device requires either IPv4 and IPv6 protocols, both simultaneously deployed, or the implementation of additional technical transition or translation measures. Transitioning to IPv6 requires software updates and address reconfiguration, so this necessitates new training and technical skills. But the need to concurrently support both IPv4 and IPv6, likely to coexist indefinitely, presents a greater impediment to those implementing IPv6. The two dominant transition techniques for supporting both IPv4 and IPv6 are called dual stack protocols and tunneling.[100]

**Dual Stack Transition**
The most common transition mechanism is called a dual stack approach, which essentially requires running both IPv4 and IPv6 simultaneously, as shown in figure 4.2. Also called dual IP layer, the dual stack option

100. Robert Gilligan and Erik Nordmark, "Transition Mechanisms for IPv6 Hosts and Routers," RFC 2893, August 2000.

**Figure 4.2**
Dual stack IPv6 transition

involves the implementation of both IPv4 and IPv6 protocols within routers, servers, firewalls, end devices, and other network components. This is the prevalent approach for upgrading to IPv6 but it has significant drawbacks. First, having to support both protocols adds great complexity, requiring two addressing plans and more complicated network management requirements. Second, implementing a dual protocol network has costs, including implementation costs, ongoing management expenditures, and personnel costs, and also requires additional system resources. Third, it can safely be assumed that there will be IPv4-only environments for the foreseeable future. These environments will not have dual IPv4 and IPv6 protocols simultaneously implemented. The problem is that new IPv6 deployments, in order to communicate with these legacy implementations will have to implement legacy capabilities, meaning IPv4. This approach requires IPv4 addresses so it has the significant drawback of not directly solving the problem IPv6 was designed to address: that of IPv4 address scarcity. In other words, the dual stack approach does not solve the underlying problem of Internet address depletion because it still requires IPv4 addresses.

## Tunneling

An alternative technique, tunneling, would encapsulate (embed) packets of IPv6 information within IPv4 packets for transmission over an IPv4 network or, inversely, encapsulate IPv4 packets within IPv6 packets before traversing an IPv6 network. Because of the predominance of IPv4 networks, must tunneling approaches have involved the tunneling of IPv6 traffic over IPv4 networks, as shown in figure 4.3. In practice, there are many forms of tunneling. For example, router-to-router tunneling involves the transmission of IPv6 packets between IPv4/v6 routers over an IPv4 infrastructure. Host-to-host tunneling involves the encapsulation and transmission of IPv6 packets over an IPv4 infrastructure between IPv4/v6 devices.

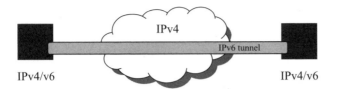

**Figure 4.3**
Host-to-host IPv6 tunneling

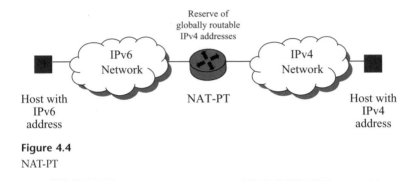

**Figure 4.4**
NAT-PT

### Translation

There is a third option. One of the rationales for the development of IPv6 was for end-to-end IPv6 to replace the prevailing address conservation approach of using IPv4 NATs. As mentioned earlier, many in the Internet's technical community frowned upon NAT approaches because they violated the end-to-end architectural principle of the Internet. It appears that a possible solution for enabling interoperability between IPv4 and IPv6 networks is a variation of this maligned translation approach. Dual stack and tunneling approaches have historically been the more common techniques for introducing IPv6 into networks, but translation is likely to become increasingly prevalent in dealing with mixed IPv4/IPv6 networks around the globe. One translation variation designed to allow IPv4 and IPv6 end nodes to communicate is to use a combination of protocol translation and address translation. This approach, called NAT-PT[101] (network address translation–protocol translation), does not require tunneling or dual stack protocol implementations. Under this scenario, depicted in figure 4.4, devices natively using IPv6 can communicate with computing devices

101. See, for example, George Tsirtsis and Pyda Srisuresh, "Network Address Translation—Protocol Translation (NAT-PT)," RFC 2766, February 2000.

using IPv4 if the information they exchange passes through a NAT-PT device. This intermediary translation device holds a reserve of IPv4 addresses, which it can dynamically assign to IPv6 devices.

The translation device performs two functions. In addition to translating Internet addresses, a NAT-PT device also translates between IPv4 and IPv6 packet headers.[102] This approach, technically, is promising because it is completely invisible to end users. But it has the architectural implication of further eliminating the end-to-end architectural approach, which carries its own risks such as somewhat complicating network-layer security services and providing, through central and concentrated NAT-PT locations, control points that could be used (e.g., by repressive governments) for information surveillance, filtering, or censorship.

### Transition Prospects

Not surprisingly, the issue of upgrading to IPv6 has been an impassioned topic within the Internet's technical community. At a March 2007 informal gathering of Internet service operators on the day preceding the 2007 Chicago IETF meeting, Randy Bush of the Internet Initiative Japan delivered a bleak presentation about IPv6 transition prospects.[103] Bush described the IPv6 situation as being "designed with no serious thought to operational transition," that the transition problems could have been avoided if IPv6 had variable length addressing rather than 128-bit fixed addressing, and that "there are no simple, useful, scalable translation or transition mechanisms."[104]

The Internet technical community has spent a great deal of time concerned about Internet address conservation and transition techniques. IETF participant John Curran published, in July 2008 as an informational RFC, an Internet-wide transition plan to IPv6.[105] The underlying objective

102. See Erik Nordmark, "Stateless IP/ICMP Translation Alogorithm (SIIT)," RFC 2765, February 2000.

103. The Internet Engineering and Planning Group (IEPG) is a group made up primarily of Internet service providers/operators, with the goal not to develop technology but to promote "a technically coordinated operational environment of the global Internet." The IEPG informally gathers the Sunday preceding IETF meetings.

104. Randy Bush, "IPv6 Transition and Operational Realities," IEPG-Chicago, July 2007. Accessed at http://www.iepg.org/2007-07-ietf69/070722.v6-op-reality.pdf.

105. John Curran, "An Internet Transition Plan," (Informational) RFC 5211, July 2008.

of Curran's recommendation, and of the overall Internet technical community, is the achievement of an eventual transition from a predominantly IPv4 to a predominantly IPv6 global Internet environment. Curran's transition plan, which is quite succinct, acknowledges that specifying changes that every single system connected to the Internet *must* undergo is unreasonable and implausible. Curran's recommended plan involves a three-phase transition involving a preparatory stage, a transitional stage, and a post-transitional stage. The fact that the Internet community acknowledged, in 2008, that the transition to IPv6 was still in a "preparatory" stage is a starkly dissonant reality from the government transition plans that, back in 2000, sought widespread IPv6 adoption by 2005.

Curran's three recommended phases for facilitating global IPv6 migration can be summarized as follows: In the preparatory phase, service providers should offer pilot IPv6 services; organizations with public-facing Internet servers (e.g., web and email servers) *should* add IPv6 capability; and organizations *may* offer IPv6 services within internal networks. In the transitional phase, which Curran described as lasting until December 2011, service providers *must* offer IPv6 services to customers, whether IPv6-only services or IPv6 through one of the transitional mechanisms described above; organizations with public-facing servers *must* provide IPv6 for these servers; and organizations *should* use internal network IPv6. In the post-transition phase, which Curran describes as beginning in 2012 and extending into the future, service providers *must* offer IPv6 services, which *should* be native IPv6; all public facing Internet servers (e.g., websites) *must* implement IPv6; and organizations *should* provide internal IPv6 connectivity. Even in this post-transition scenario, Curran states that service providers *may* continue to offer IPv4 Internet services and that organizations *may* continue to use IPv4 internally.

If organizations may still use IPv4 internally and service providers may still offer IPv4 Internet services, the Internet standards community is acknowledging that it expects IPv4 to exist in Internet infrastructures indefinitely. So the question of incentive is an open issue, and a classic collective action problem. If an organization has sufficient Internet addresses and has complete Internet functionality, what would provide the motivation to upgrade to IPv6? The critical mass of upgrades would have to be for the public good rather than for the good of the organization. A necessary precursor for the transition to IPv6 is the desire among individual organizations to act in the common good and to consider connectivity as its own good, as Curran describes:

[T]he requirement for existing Internet-connected organizations to add IPv6 connectivity (even to a small number of systems) will be a significant hurdle and require
a level of effort that may not be achievable given the lack of compelling additional
benefits to these organizations [RFC1669]. This transition plan presumes that "connectivity is its own reward" [RFC1958] and that there still exists a sufficient level of
cooperation among Internet participants to make this evolution possible.[106]

Organizations with ample addresses do not, without a critical need or a
new application that requires IPv6, have incentive to upgrade. But, as this
chapter described, governments have expressed a variety of public incentives to upgrade. Despite a decade of government adoption strategies, the
deployment of IPv6 has been slow. There are three forces working against
widespread protocol adoption: the conservative momentum of existing
protocols, the absence of free market demand for protocols, and the decentralized nature of infrastructural control over the Internet.

Historian of technology Ken Alder has argued: "if standards are a matter
of political will as much as of economic or technical readiness, then reaching an agreement on standards depends as much on myths as on science,
especially on myths *about* science."[107] IPv6 is a routing and addressing
specification, not a specific application, but advocates have espoused
buoyant expectations about IPv6 spreading democratic freedoms, thwarting unemployment, and enabling distributed warfare. One myth is the
claim of IPv6 as self-evidently more secure than IPv4 and the use of this
as an apologia for upgrading. Advocacy groups, national government technology councils, the technical media, and networking vendors promoted
IPv6 as self-evidently more secure than IPv4, but in practice, protocol
vulnerability reports from CERT, GAO technical assessments, and security
experts cast doubt on these claims. Implementation realities are more
complicated than paper specifications or high-level strategies.

Any magnified IPv6 claims have not diminished the underlying concerns
about inherent resource constraints, distribution inequities, or projected
address requirements for emerging applications. This chapter has described
how most of those driving IPv6 adoption have abrogated laissez-faire
approaches, instead delivering top-down mandates such as Japan's national
IPv6 directive or the DoD's IPv6 pronouncement. With the exception
of the US Commerce Department's positions, state interventions have
selected the technology, IPv6 products, that vendors must develop

106. Ibid.
107. Ken Alder, *The Measure of All Things: The Seven-Year Odyssey and Hidden Error
That Transformed the World*, New York: Free Press, 2002, p. 327.

rather than advocating for competitive markets to drive Internet product adoption.

Finally, national strategies suggest an important general theme about Internet protocol adoption: Internet protocol strategies reflect competitive struggles for control of the Internet and for economic dominance in the Internet industry, and reflect how protocols, or even talk about protocols, can bolster and reinforce political objectives. Distinct from resource requirements, governments selected IPv6 as a new arena in which market hegemony had not yet been established. Conversely, the conservative position of maintaining the status quo by deflecting federal standards involvement onto market mechanisms sought to maintain the dominance of those with ample addresses, resource control, or market leadership in Internet products. The promise of IPv6 aligned with a variety of political objectives: a homogenizing specification advancing European unification and economic competitiveness; governmental promises of IPv6 thwarting economic stagnation in Japan or unemployment in Korea; the DoD promise of IPv6 for a secure and distributed war on terrorism; or the potential for the United States to subvert economic threats from India and China. In most cases the issue of address space exhaustion existed as a tangential rationale. In general, political and technical objectives were mutually cast as unquestioned certainties, with the concealed complexity of the IPv6 specification all but precluding public ability to question the efficacy of the standard to achieve promised objectives.

# 5   The Internet Address Space

The Information Age will continue to create new artifacts, some that carry great value. We should not stand idly by and let rights to the assets of this new Age be determined haphazardly, thereby almost certainly guaranteeing that they go to people in the best position to take quick advantage of them. We should try to analyze them thoughtfully, remembering our real-world experience with inequality and exploitation and trying not to recreate it in new worlds.[1]

—Anupam Chander, "The New, New Property"

Internet addresses are the new artifacts of the information age. They are the finite resources necessary for being online. The design of the Internet's underlying architecture dictates that each virtual address used to route information over a network must be globally unique. Maintaining this global uniqueness has required centralized oversight of the finite pool of IP addresses so that duplicate addresses are not concurrently used. This address management function is one of the most centralized of Internet governance functions. While coordination is centralized, IP addresses are completely virtual, not necessarily tied to geographical location, and universal. Unless specifically filtered out by a firewall or other technological intermediary, the standard design of Internet addresses, theoretically, allows them to reach any part of the Internet regardless of nation, jurisdiction, architectural ownership, or geographical vicinity.

This combination of criticality, centralized control, and scarcity raises many Internet governance questions. The first question is *who* controls these addresses (and who should control these addresses)? If these resources are central to the ability of citizens to use the Internet and of nations to participate in the global knowledge economy, a related concern is the basis upon which any authority has legitimacy to oversee these finite resources.

1. Anupam Chander, "The New, New Property," 81 *Texas Law Review* 715–97 (2003).

In addition to questions about jurisdictional control and legitimacy, another Internet governance concern is *how* these finite resources are distributed and whether directed toward global fairness, market efficiency, rewarding first movers, or other objective. How these resources are distributed determines whether there is equal opportunity online or perpetuation of inequality in the distribution of access and therefore the distribution of wealth. The mode of distribution of finite resources and the rules regarding their allocation should not at all be viewed as natural or technically fixed, but as socially and institutionally constructed. Distribution can occur in any number of ways—market-based approaches, through government control or regulation, through community-based distribution, or through private institutional control. The prospect of geographical, political, or socioeconomic disparities in IP address allocation raises questions about equitable control, distribution, and possession of technologically generated resources within a system that transcends national boundaries.

Another governance concern addresses *sufficiency*, the question of whether there exist adequate resources to meet current and expected global demand.

Despite the criticality of these Internet governance questions, most public controversies and scholarship about Internet resource control have concentrated on domain names, the human readable text strings (e.g., www.yale.edu or www.yahoo.com) associated with IP addresses. The allocation of alphanumeric domain names raises questions of distributive equality but also frequently encounters more readily identifiable questions about antitrust law, intellectual property, free speech, and cultural standards of decency. Many policy questions about domain names involve trademark concerns: who should own www.united.com, United Airlines, United Arab Emirates, or United Van Lines? Other issues involve tensions among free speech, decency, and censorship: who decides what domain name is controversial or objectionable?

IP addresses have not historically received proportionate attention, partly because users do not directly engage Internet addresses. They are invisible. Addresses raise a distinct set of governance questions. The availability of IP addresses speaks to the most fundamental questions about who can access the Internet, how they access the Internet, and whether sufficient resources exist for equitable participation in the information society.

This chapter seeks to elevate the issue of IP address space design, allocation, and control as a critical Internet governance question involving issues of institutional control, national jurisdiction, and global access to knowledge. It historically traces the progression of the Internet address space from its 1960s inception, to the development of the IPv4 address space, to anticipation of potential Internet address space exhaustion by the Internet's technical community in 1990, to the new IPv6 address structure. The chapter includes accounts of dissenting arguments challenging predictions of Internet address scarcity and also describes intractable governance dilemmas involving international and nongovernmental struggles for control of Internet addresses.

## Internet Resources circa 1969

New technologies create new, technologically derived resources. Radio systems introduced the electromagnetic spectrum's radio frequency band. The Internet created unique binary addresses. Unlike electromagnetic spectrum (which includes harmful ultraviolet, X-ray, and gamma-ray bands), no natural trait constricts the number of theoretically possible Internet addresses. The Internet standards community established specifications dictating the length of Internet addresses and therefore the number of devices able to interconnect.

The topic of network addresses appeared in the premiere Request for Comment, RFC 1, "Host Software." UCLA's Stephen Crocker authored RFC 1 in 1969, several months before the UCLA ARPANET node, the first of four original ARPANET nodes, became operational and prior to any definitive decisions about the applications the network would eventually support. RFC 1 enumerated tentative specifications for the Interface Message Processor (IMP) software and host-to-host connections. ARPANET researchers decided to allocate 5 bits to information headers as a destination code for the IMPs.[2] The allocation of 5 bits as a destination address would have theoretically provided $2^5$, or 32, unique destination codes:

```
00000  00100  01000  01100  10000  10100  11000  11100
00001  00101  01001  01101  10001  10101  11001  11101
00010  00110  01010  01110  10010  10110  11010  11110
00011  00111  01011  01111  10011  10111  11011  11111
```

Expanding the total number of addresses would require expanding the size of the binary code. Each additional bit would double the number of available addresses. For example, increasing the binary code to 6 bits would provision $2^6$, or 64 addresses; increasing the binary code to 7 bits

2. Steve Crocker, "Host Software," RFC 1, April 1969.

would expand the number of unique addresses to $2^7$, or 128; and increasing the code to 8 bits would provide $2^8$, or 256 unique addresses, and so forth.

The researchers gradually augmented the number of addresses as they anticipated requirements for connecting more devices. In 1972 Internet engineers extended the address size to 8 bits, increasing the number of possible device connections to $2^8$, or 256. In 1976, seven years after the 1969 operational installation of IMP 1 at UCLA, the ARPANET interconnected 63 hosts. The 256 available destination codes more than sufficed to connect these devices. A gradual ARPANET expansion occurred within a mid-1970s computing context dominated by mainframe computers, with a modest minicomputer industry, but prior to widespread availability of personal computers. In this experimental environment in which expensive mainframe computers predominated, widespread growth or even success of the ARPANET was not inevitable. Even if successful, as Katie Hafner and Matthew Lyon explained in *Where Wizards Stay up Late: The Origins of the Internet*, "Who but a few government bureaucrats or computer scientists would ever use a computer network?"[3]

As Janet Abbate explains, a phenomenon unforeseen by ARPANET developers was the emergence of electronic mail in the 1970s as the network's most widespread and expansive application. Prior to the network's development, ARPANET Project Manager Larry Roberts downplayed electronic messaging as a possible application, focusing instead on resource sharing and file transfer.[4] But rather than primarily interconnecting computing resources as anticipated, ARPANET users developed and embraced programs and protocols for real-time messaging that supported collaborative work and served as a communication forum for the growing ARPANET community. The unanticipated application of electronic mail continued to interest users.

Electronic mailing lists became both a driver of increased network usage and a reflection of the ARPANET's growing role as a communications platform for a rapidly expanding electronic community. Rather than providing communications between two computers, mailing lists enabled large groups of people with common interests and identities to communicate in an open forum. Mailing lists contributed to the unexpected growth in the size of the network, played an important role in facilitating

3. Katie Hafner and Matthew Lyon. *Where Wizards Stay up Late: The Origins of the Internet*, New York: Simon and Schuster, 1996, p. 104.
4. Janet Abbate, *Inventing the Internet*, Cambridge: MIT Press, 1999, pp. 106–10.

communications among Internet standards and technology communities, and reflected shared values of open communications and collaborative development within the Internet user/developer culture.

RFC 791 (1981) introduced the Internet Protocol standard, later called IPv4, expanding the size of each IP address to a 32-bit code divided into a network prefix and a host prefix. Mathematically this binary address size of 32 bits would support more than four billion hosts, calculated as $2^{32}$, or roughly 4.3 billion. As described earlier, each of the more than four billion unique addresses under the IPv4 standard was simply a combination of 32 0s and 1s such as 00011110000101011100001111011101, or 30.21.195.221 in conventional shorthand notation. Four billion plus addresses seemed immense at the time but still required centralized coordination and distribution to guarantee global uniqueness for each address.

## Distributing Limited Resources

If each device connected to the Internet required a globally unique address from the pool of almost 4.3 billion IPv4 addresses, some mechanism would have to provide central administration, tracking, and distribution of addresses. Jon Postel performed this function for years. As casually noted in the RFCs documenting assigned Internet numbers in the 1970s and into the early 1980s, "The assignment of numbers is also handled by Jon."[5]

Number assignment in the context of the 1970s and 1980s was hardly controversial work. Postel worked at the University of Southern California's (USC) Information Sciences Institute (ISI), then a US Department of Defense funded institution. Postel's activities were DARPA-sanctioned, and this association, along with his technical expertise, provided legitimacy for him to act as a central authority distributing addresses to what were then primarily American institutions. Within the Internet's technical community, Postel had considerable stature as a respected insider and early ARPANET contributor. In addition to technical credibility, experience, and DARPA-sanctioned legitimacy, Postel had direct personal ties with others prominently involved in ARPANET development. Postel and Vinton Cerf had attended Van Nuys High School together in California's San Fernando Valley and were both UCLA graduate students working for Leonard Kleinrock on the ARPANET project beginning in the late 1960s.

5. Jon Postel, "Assigned Numbers," RFC 739, November 1977.

Cerf later memorialized Postel as the Internet's "North Star,"[6] and recalled, "Someone had to keep track of all the protocols, the identifiers, networks and addresses and ultimately the names of all the things in the networked universe. And someone had to keep track of all the information that erupted with volcanic force from the intensity of the debates and discussions and endless invention that has continued unabated for 30 years."[7] Postel was referred to as a "rock," an "icon," and a "leader" and was described as "our Internet Assigned Numbers Authority."[8]

Joyce Reynolds, also at USC's Information Sciences Institute, a major contributor to the Internet RFC process, and author of numerous RFCs, assumed additional day-to-day address assignment responsibility in 1983.[9] Cerf described Reynolds and Postel as functioning "in unison like a matched pair of superconducting electrons—and superconductors they were of the RFC series. For all practical purposes, it was impossible to tell which of the two had edited any particular RFC."[10] From 1983 through 1987 the network assignment RFCs instructed those wanting network numbers to "please contact Joyce to receive a number assignment."[11] The functions performed by Postel and Reynolds at the USC-ISI were called the "Internet Assigned Numbers Authority" or IANA. Institutions freely obtained addresses on an as-requested basis. The primary purpose of central address distribution was to ensure the global uniqueness of each address. In the 1970s and 1980s there were ample addresses and the possibility of exhausting the Internet address space seemed almost inconceivable.

As Internet growth expanded in the late 1980s, number assignment responsibility institutionally shifted to a more formal government-funded structure, the Defense Data Network–Network Information Center (DDN-NIC), sponsored by the US Defense Communications Agency[12] and operated at Stanford Research Institute (SRI). Milton Mueller suggests that this shifting of assignment authority followed a Defense Department pattern.

---

6. Vinton Cerf quoted in Internet Society Press Release, "Internet Society Statement on the Death of Jon Postel," Reston, VA, October 1998.
7. Vinton Cerf, "I remember IANA," (Informational) RFC 2468, October 1998.
8. Ibid.
9. Joyce Reynolds and Jon Postel, "Assigned Numbers," RFC 870, October 1983.
10. From Vinton Cerf's entry in "30 Years of RFCs," RFC 2555, April 1999.
11. See RFC 870 (1983), RFC 900 (1984), RFC 923 (1984), RFC 943 (1985), RFC 960 (1985), RFC 990 (1986), and RFC 1010 (1987).
12. Sue Romano, Mary Stahl, Mimi Recker, "Internet Numbers," RFC 1020, November 1987.

As technological systems transfer from experimental to operational, authority shifts from researchers to a military agency.[13] The DDN-NIC did distribute addresses, but, as Cerf described in RFC 1174, IANA, meaning primarily Postel, retained responsibility and had "the discretionary authority to delegate portions of this responsibility."[14] In other words, the DDN-NIC would handle requests and provide address (and name) registration services but Postel still controlled the allocation of addresses to the NIC for further allocation or assignment.

The easiest way to explain this is to differentiate between allocation and assignment. Although the terms are sometimes used interchangeably, to *allocate* address space means to delegate a block of addresses to an entity for subsequent distribution to another entity. To *assign* address space means to distribute it to a single entity, such as a corporation, for actual use. The centralized entity of IANA allocated large address blocks to registry organizations like the DDN-NIC to either assign directly to end users or to allocate to ISPs for assignment to end users. This distinction between responsibility for delegating allotments of addresses to registries and the actual assignment of addresses would endure indefinitely as the DDN-NIC later transformed into the less military oriented InterNIC, which eventually transformed into the American Registry for Internet Numbers (ARIN) and various international Internet registries. A variety of entities performed address assignment, but more than anyone else, Jon Postel controlled address allocations. A colleague later eulogizing Jon Postel said, "I find it funny to read in the papers that Jon was the director of IANA. Jon was IANA."[15]

Address distribution has historically occurred outside of traditional market mechanisms of supply and demand. Mueller has enumerated five possible methods of distributing Internet resources, including the resources of Internet names and numbers:

- First come–first served
- Administrative fees
- Market pricing

13. Milton Mueller, *Ruling the Root: Internet Governance and the Taming of Cyberspace*, Cambridge: MIT Press, 2002, p. 82.
14. Vinton Cerf, "IAB Recommended Policy on Distributing Internet Identifier Assignment and IAB Recommended Policy Change to Internet 'Connected' Status," RFC 1174, August 1990.
15. Danny Cohen, "Working with Jon, Tribute delivered at UCLA, October 30, 1998," RFC 2441, November 1998.

- Administrative rules
- Merit distribution.[16]

First come–first served describes early entrants acquiring whatever resources they request or claim. The administrative fees approach, often in conjunction with first come–first served, imposes a price on resources to prevent massive hoarding of finite resources. Allocation based on market pricing allows price to reflect demand, the economic value of the resource, and the extent to which the resource is scarce. Using this method, those wanting IP addresses would purchase the quantity they required at market price. Allocative approaches could also impose administrative rules to ration resources, such as imposing a maximum allowable per user allocation or requiring organizations to demonstrate need prior to allocation. Finally, merit distribution, somewhat of a subset of the administrative rules approach, would allocate resources based on subjective merit assessments.

In the initial two decades of address distribution, addresses were received on a first come-first served basis, also called a first possession approach. Anupam Chander describes the inherent problems that can result from the approach of first possession. This allocative method appears on the surface equitable because any entity can theoretically request Internet addresses (or domain names) regardless of distinguishing characteristics such as geographical location, ethnic or national affiliation, or any other differentiation. But Chander notes that "[f]irst possession is less a theoretical justification for the distribution of private property than an assertion of power"[17] in that theoretical equality does not translate well to substantive equality. First come–first served provides distinct advantages to those already advantaged—with technology, capital, and even the knowledge itself that requesting resources is necessary. "A formally equal system may in fact play into the hands of some at the expense of others."[18]

In the case of Internet address distribution, administrative and technical decisions, as well as the existing user context, determined several additional circumstances that would ultimately contribute to the possibility of address space exhaustion:

1. addresses would be allocated in large blocks;
2. once distributed, recipient organizations would irrevocably possess these resources;

---

16. Milton Mueller, *Ruling the Root: Internet Governance and the Taming of Cyberspace*, Cambridge: MIT Press, 2002, pp. 24–25.
17. Chander, p. 733.
18. Chander, p. 734.

3. the resources would be completely free (until 1997 when US subsidization of the assignment function ceased and registries introduced minimal fees); and

4. large American research institutions and corporations requesting addresses would receive an asymmetrically large quantity of addresses relative to demand for Internet connectivity.

The following sections describe how emerging IP address constraints were not purely a mathematical limitation relative to demand but a circumstance influenced by institutional decisions about an Internet "class system" and based on large, irrevocable allocations of addresses to those American institutions involved as early users and developers.

### Initial Internet Address Constraints

Mathematically, IPv4 provided roughly 4.3 billion addresses, but several administrative and technical decisions about the composition and distribution of addresses constrained the actual number of available addresses and therefore the number of devices able to connect to the Internet.

### The Internet Class System

For reasons of technical efficiency, the IPv4 specification defined a 32-bit address as consisting of two distinct domains, a network prefix and a host number.[19] The first address segment, the network prefix, would represent the network to which a computing device was connected. The second part, the host number, would identify a specific computing device, called a "host" in 1980s network parlance. For example, the first 16 bits of an Internet address could designate a specific network, and the final 16 bits could represent various hosts on that network. IANA would provide a unique network number to an Internet user institution, which would then discretionarily assign the host numbers associated with that network number to devices on its network. This hierarchical concept did not significantly differ from the conventionally layered approach of postal addresses. A typical street address contains a six-layer hierarchy: country, zip code, state, city, street, and house number. This hierarchical structure simplifies the routing process. Intermediate postal centers need only scan a zip code to determine how to route a letter. Analogously, an Internet router need only scan the network prefix to make routing decisions. Only when a postal letter or Internet packet reaches the zip code or network

19. Jon Postel, ed., "DoD Standard Internet Protocol," RFC 760, January 1980.

destination is it necessary to process local information such as street address or host IP address. Routers rely on routing tables to decide where to forward packets, and the hierarchical network/host address structures eliminated the requirement for routing tables to include every address component, conserving storage and processing resources.

This IPv4 address division into network prefix and host number underpinned the Internet class system and set constraints on how many host addresses a single institution could receive. Rather than an individual organization requesting an ad hoc number of addresses, the network/host address division necessitated that an institution receive a network prefix address accompanied by the fixed number of host addresses associated with that prefix. Internet designers anticipated that some organizations would require large blocks of host addresses while some might only require a small number. Accordingly they originally divided IPv4 address blocks into five categories: Class A, B, C, D, and E. Class D and E addresses were reserved for multicast[20] applications and experimental uses, rendering those address blocks unavailable for general user assignment.

Rather than requesting a specific number of addresses, institutions would receive a block of addresses according to whether the assignment was designated Class A, B, or C. Recalling that each IPv4 address contained a total of 32 bits, a Class A designation divided addresses into a 7-bit network prefix (within the first octet, the highest order—i.e., leftmost—bit was set to 0) and a 24-bit local, host address. This address structure would allow for a theoretical total of 128 blocks of Class A networks ($2^7$), with each network supporting approximately 16 million ($2^{24}$) computers. In other words, only 128 organizations could receive large Class A blocks of IP addresses. Another class, called Class B address blocks, would include a 14-bit network number and a 16-bit local address, with the first two bits set to 1-0. This allowed for 16,384 ($2^{14}$) Class B address blocks, each supporting approximately 65,000 ($2^{16}$) computers. Finally, organizations could also receive a Class C address assignment, which set the first three address bits to 1-1-0 and allocated a 21-bit network number and an 8-bit local address. This would theoretically allow for 2,097,152 Class C networks ($2^{21}$), each providing only 256 addresses ($2^8$).

Figure 5.1 depicts the network and host division of a Class A, B, or C address. The rationale behind this class system was that few organizations would require more addresses than a Class C address block provided. In

---

20. Multicast is the ability to transmit to all IP-addressed computers on a network or subnetwork, usually for autoconfiguration.

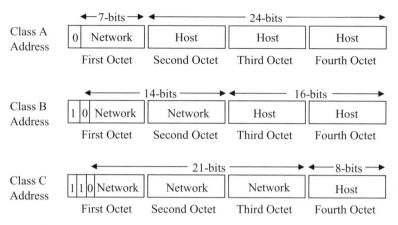

**Figure 5.1**
Network and host address divisions

**Table 5.1**
Class A, B, and C Internet addresses

| Type of address block | Number of available blocks | Number of assignable host addresses per block |
| --- | --- | --- |
| Class A | 128 | 16,777,216 addresses |
| Class B | 16,384 | 65,536 addresses |
| Class C | 2,097,152 | 256 addresses |

the 1980s context, it was not readily conceivable that many organizations would require as many as 256 addresses, so the more than two million available Class C networks seemed sufficient. RFC 1117, "Internet Numbers," documents a snapshot of the assigned Class A, B, and C Internet address assignments in the 1980s and describes the binary structure of the address classes.[21] Table 5.1 summarizes the number of available Class A, B, and C address blocks and the number of local, or host, addresses supported by each block.

The hierarchical structure and class system of Internet addresses inherently decreased the theoretical maximum number of available addresses. Protocol developer Christian Huitema[22] was among those within the

21. Sue Romano et al., "Internet Numbers," RFC 1117, August 1989.
22. Huitema has worked at CNET (Centre National d'Etudes des Telecommunications), INRIA (Institut National de Rechereche en Informatique et en Automatique),

Internet standards community who analyzed issues of maximum theoretical address availability.[23] The mathematical maximum of 4.3 billion decreased because Class D addresses were reserved for multicast applications and Class E addresses were reserved for experimental uses. The number of reserved Class D and E addresses totaled 536,870,912. Eliminating these addresses from the theoretical maximum reduced the number of available addresses from roughly 4.3 billion to less than 3.8 billion. Two entire Class A address blocks, 0 (null network) and 127 (loopback) were made unavailable for general allocation, eliminating 33,554,432 additional addresses from allocation availability. Decisions about allocating class resources created this diminishment of available addresses, but the real impact of the class system was that it ensured the allocation of often unnecessarily enormous blocks of addresses to some institutions. The following section will discuss how many of these institutions did not require or use the majority of addresses allocated to them. In other words, these allocated addresses were unused yet rendered unavailable for distribution to others.

### Address Assignment Asymmetry

Address assignment asymmetry significantly constrained the available IP address space. The class system allowed for assigning more than 2,000,000 organizations Class C address blocks with 256 addresses each. By the late 1980s many institutions did not yet require 256 addresses but anticipated that they would at some future time. A tendency among organizations was to request Class B address blocks providing 65,536 IP addresses rather than a small Class C address block of 256 IP addresses. The term "hoarding" is not appropriate, but this planning for future growth resulted in organizations using a relatively small number of their Class B addresses and leaving the rest unused, yet unavailable for other users. If an organization with a Class B assignment actively used 1,000 Internet addresses, 64,536 addresses would remain dormant yet unavailable for others to use.

A much greater allocative inefficiency ensued among institutions with Class A allocations. Even a large corporation connecting a then-exorbitant 10,000 devices to the Internet would result in 16,767,216 addresses unused and unavailable. Rather than requesting an ad hoc number of addresses supporting current requirements and anticipating future growth, such as

---

Bellcore, and Microsoft and has been a member of the Internet Architecture Board and Internet Society.

23. Christian Huitema, "The H Ratio for Address Assignment Efficiency," RFC 1715, November 1994.

**Table 5.2**
Internet host statistics, 1981 to 1989

| Year | Number of Internet hosts |
| --- | --- |
| 1981 | 213 |
| 1982 | 235 |
| 1983 | 562 |
| 1984 | 1,024 |
| 1985 | 1,961 |
| 1986 | 5,089 |
| 1987 | 28,174 |
| 1988 | 56,000 |
| 1989 | 159,000 |

30,000 addresses, organizations would have to request a small block of 256, a large block of more than 65,000, or an enormous block of more than 16 million addresses. The technical rationale for the Internet class system was consideration of router table sizes, but built into the structural characteristics of the Internet class system was the potential for allocative inefficiency and stockpiling of surplus addresses.

The historical relationship between the number of addresses distributed and the number of addresses actually used demonstrates this inefficiency. In 1981, according to Stanford Research Institute's statistics immortalized in the RFC system, the Internet supported 213 hosts (devices). Table 5.2 provides a snapshot of the Internet's scope during the 1980s.[24]

The majority of hosts used a single IP address (though some had multiple IP addresses), so the table provides an approximate, though underestimated, indication of the demand for IP addresses during the 1980s. What was the relationship between the number of hosts connected by the Internet and the number of addresses already assigned? At the time, SRI's NIC maintained statistics about both the number of Internet hosts and the number of assigned addresses.

If the Internet connected 159,000 hosts in 1989, as reported, and if most of these hosts required a single unique IP address, then at least 159,000 addresses should have been allocated at that time. According to 1989 NIC records,[25] large universities, defense agencies, and corporations already

24. Statistics on the number of Internet hosts from "Internet Growth (1981–1991)," RFC 1296, January 1992.
25. Sue Romano, Mary Stahl, and Mimi Recker, "Internet Numbers," RFC 1117, August 1989.

held 33 Class A address blocks, 1,500 Class B address blocks, and numerous Class C addresses. The assigned Class A address assignments alone expended more than 500 million Internet addresses. The Class B assignments exhausted a pool of more than 100 million.

In other words, in 1989 there were very roughly 159,000 Internet hosts and more than 600 million addresses assigned, or a ratio of almost 4,000 addresses assigned per Internet host. A substantial reason for this unbalanced address to host ratio, as mentioned, was the structural design of the Class A, B, and C address blocks, intended to save router processing requirements but mathematically exhausting large, unused blocks of IP addresses.

Another explanation for some of this high address to host ratio was that in the late 1980s many corporations operated private TCP/IP networks disjoint from the broader public Internet. These networks required IP addresses. Institutions operating private TCP/IP networks could have implemented any IP numbering scheme, as long as the numbers were unique within each private network environment, but corporations frequently sought globally unique IANA assignments, presaging a future interconnection of their private TCP/IP networks to a public network or to other private TCP/IP networks operated by business partners, customers, or suppliers. Using these globally unique, assigned addresses would allow corporations later connecting to the public Internet to avoid the cumbersome task of renumbering networks.

### Initial Allocation to US Institutions

The principal recipients of Internet addresses in the 1970s and 1980s were American institutions: universities, government agencies, corporations, and military networks. The RFCs divided address recipients into four categories: research, government agency, commercial, and military. Not surprisingly, many holders of large Class A address blocks were organizations involved in the early development and use of ARPANET technologies, such as BBN, UCLA, Stanford, and a variety of defense agencies. By the late 1980s, the large address holders expanded to include then-dominant technology corporations like IBM, DEC, HP, and Xerox; prominent universities; and a variety of defense and governmental agencies and commercial networks.

The following are some of the institutions which, in 1989, held 16 million or more Internet addresses:

- AT&T Bell Labs
- Bolt Beranek and Newman

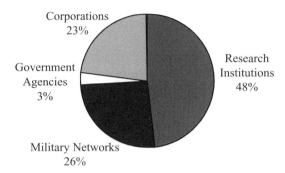

**Figure 5.2**
Address distribution by institution type

- DoD Intel Information Systems
- Defense Data Network
- General Electric Company
- Hewlett-Packard Company
- International Business Machines

A few institutions from Great Britain, Canada, and Japan held Class A address blocks by the late 1980s, but the vast majority of address holders were American. Among the addresses already distributed by 1990, approximately 80 percent were held by government, military, and research institutions and roughly 20 percent were held by American corporations.[26] Figure 5.2 illustrates the 1990 address distribution by type of institution. Figure 5.3 provides a more detailed snapshot of the address distribution, derived from raw numbers published in RFC 1166 (July 1990), delineated by address class and institution type.

Figure 5.3 illustrates several characteristics of relatively early IP address distribution. First, the majority of assigned addresses were part of large, Class A address blocks, many distributed in the 1970s and 1980s to institutions involved in early Internet use and development. Second, research institutions, government agencies, and military networks received the bulk of address allocations. Corporations controlled only 23 percent of address assignments, and many of these were for private TCP/IP networks rather than public Internet connectivity.

Finally, the numerical data prefigured a problem that would later surface: a shortage of unassigned Class A and B address blocks. Comparing these

---

26. Percentages calculated from the address allocation data published in RFC 1166: Sue Kirkpatrick et al., "Internet Numbers," RFC 1166, July 1990.

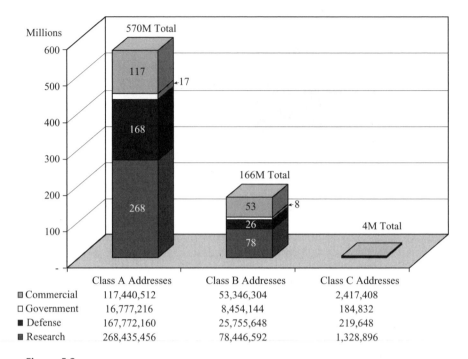

| | Class A Addresses | Class B Addresses | Class C Addresses |
|---|---|---|---|
| ▨ Commercial | 117,440,512 | 53,346,304 | 2,417,408 |
| ▢ Government | 16,777,216 | 8,454,144 | 184,832 |
| ▪ Defense | 167,772,160 | 25,755,648 | 219,648 |
| ▪ Research | 268,435,456 | 78,446,592 | 1,328,896 |

**Figure 5.3**
1990 Internet address distribution

1990 data from RFC 1166 with the theoretical maximum number of Class A, B, and C addresses in 1990, fewer than 1 percent of Class C addresses were distributed but 27 percent of Class A addresses and 15 percent of Class B addresses were already assigned. The Internet had experienced rapid growth by the close of the 1980s but clearly supported relatively few hosts relative to the number of Internet addresses already assigned. Despite the relatively small number of hosts, institutions held more than 600 million addresses—all prior to the World Wide Web, rapid international growth, home Internet access, and widespread corporate connectivity to the public Internet.

Those promoting IPv6 have consistently invoked historical address allocation inequities between US institutions and those of other countries as underlying rationales for upgrading. This section has described the context of these address distribution asymmetries but it is also interesting how false narratives about this history have become an indelible part of the IPv6 discourse. One highly reproduced description of address inequity has noted that Stanford University controls more IP addresses than the People's

Republic of China. Stanford University was one of the institutions appor-
tioned a Class A block of more than 16 million Internet addresses prior to
1980.[27] In addition to its Class A assignment, Stanford also controlled four
Class B networks, providing approximately 250,000 addresses.

In the late 1990s, however, Stanford voluntarily relinquished its 16
million plus Class A addresses to IANA and completed a renumbering of
its network addresses[28] by mid 2000. This renumbering process required a
laborious conversion of more than 50,000 network devices from numbers
within its Class A allocation to numbers from its four Class B network
address blocks. Prior to 2000, China held the equivalent of a Class A
address block, or 16,777,214 addresses, indeed fewer than Stanford controlled before its decision to voluntarily relinquish addresses. China steadily
requested and received additional addresses in the following years. The
address distribution circumstance of Stanford University holding more IP
addresses than China has not been the case for years.

Despite this, years after Stanford relinquished addresses and China
received additional address allocations, IPv6 descriptions and advocacy in
government policy documents, at conferences, and in the press, have
routinely cited the "statistical fact" that Stanford University controls more
IP addresses than China. Thousands of statements have reproduced this
assertion. Mainstream technical journals such as *IEEE Computer* have
referred to the outdated comparison.[29] A *Business Communications Review*
column suggested that "Stanford University is assigned more IPv4 addresses
than the entire nation of China."[30] Silicon.com argued, in 2003, that
"The whole of China has for instance been allocated just nine million
global IP addresses—Stanford University alone has twice that. . . ."[31] The
Stanford and China address comparison even appeared in the 2002
Commission of the European Communities' IPv6 strategy document to
the European Parliament as proof of Internet address scarcity and as

27. Jon Postel, "Assigned Numbers," RFC 770, September 1980, p. 1.
28. Stanford University's announcement "IP Address Changes at Stanford," relinquishing its Class A address block and renumbering its network to its four Class B
networks. Accessed at http://www.stanford.edu/group/networking/NetConsult/
ipchange/index on August 1, 2005.'
29. See citation in George Lawton's "Is IPv6 Finally Gaining Ground," *IEEE Computer*, August 2001, p. 12.
30. Eric Knapf, "Whatever Happened to IPv6," *Business Communications Review*,
April 2001, pp. 14–16.
31. Simon Marshall, "Convergence: IPv6 migration–a necessary pain?" *Silicon.com*,
June 5, 2003.

further justification for the need to immediately upgrade to IPv6. This enduring narrative about Stanford possessing more addresses than China illustrates how statistical "facts" cited by technology advocates, the media, and government institutions can be incorrect or, in this case, outdated. Nevertheless, Stanford's decision to relinquish addresses was not typical. Most institutions have retained their original address assignments, and historically US institutions have controlled disproportionate percentages of IPv4 addresses.

## Address Conservation Strategies

While the Internet standards designers worked on developing the next generation Internet protocol in the 1990s, they also introduced technical measures to help conserve existing addresses, including network address translation (NAT) and classless interdomain routing (CIDR). The Internet class system for IPv4 addresses, designed in part to minimize router processing overhead, resulted in uneven address distribution patterns such as a single corporation possessing millions of addresses but only using 20,000. The IETF engineered CIDR[32] to make address assignments less profligate and to promote routing efficiency. As the IETF RFC describing the rationale for CIDR explained:

The IP address space is a scarce shared resource that must be managed for the good of the community. The managers of this resource are acting as its custodians. They have a responsibility to the community to manage it for the common good.[33]

The main contribution of CIDR was the elimination of the class address distinctions to promote more flexible and efficient allocations of IPv4 address allocations. CIDR also offered route aggregation techniques whereby a single router table entry could represent thousands of address routes. This type of aggregation reduced the number of decisions for each router, in turn reducing processing time and router table size. Each packet of information to be routed would contain a prefix length, often referred to as a *bit mask*, notifying the router of the length of the network prefix it should read. This CIDR approach enabled routers to read all bit sizes of addresses rather than only the fixed 8-bit, 16-bit, or 24-bit network numbers under the Internet class system.

32. RFCs 1517, 1518, and 1519 document the classless interdomain routing approaches.
33. Yakov Rekhter and Tony Li, "An Architecture for IP Address Allocation with CIDR," RFC 1518, September 1993.

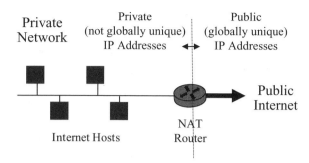

**Figure 5.4**
Network address translation

In addition to CIDR, the IETF introduced address translation to stave off potential resource depletion.[34] As mentioned earlier, NAT techniques allow a network device, such as a router, to employ a limited number of public IP addresses to mediate between the public Internet and a private network with many unregistered (fabricated) IP addresses; see figure 5.4. As an oversimplified example, a single publicly unique address could serve a local area network of twenty computers. When a computer on a private network accesses the public Internet, the NAT device dynamically allocates a globally unique, temporary IP address. When the same computer communicates with devices within the private network, it uses a private, non–globally unique address. Address translation conserves addresses by allowing numerous devices to share public IP addresses. The technique has also enabled some institutions with a large installed base of private IP addresses to connect to the Internet without laboriously converting entire networks from private (not IANA assigned) addresses to public IP addresses.

Despite its origination in the IETF, many in the Internet's standards-setting community have criticized increased NAT usage because it violates the end-to-end architectural philosophy which has underpinned the Internet (and precursor networks) since its inception. Internet engineers first articulated this philosophy in the mid-1980s[35] and later formalized this

34. See, for example, Kjeld Egevang and Paul Francis, "The IP Network Address Translator," RFC 1631, May 1994.
35. An articulation of the end-to-end architectural philosophy appears in two mid-1980s papers: John Saltzer et al., "End-to-End Arguments in System Design," 2 *ACM TOCS*, November 1984, pp. 277–88; and Dave Clark, "The Design Philosophy of the DARPA Internet Protocols," *Proceedings of SIGCOMM 88, ACM COR*, vol. 18, August 1988, pp. 106–14.

Internet principle in the IAB's "Architectural Principles of the Internet" document.[36] The architectural principle responded to a design question about where to place intelligent functions within a communications network. Some of these functions included congestion control, error detection and correction, encryption, and delivery confirmation. Internet engineers in the 1980s decided these functions should reside at network end points rather than at points within the network. Under this design philosophy, network routers would efficiently forward packets to their destinations with other functionality performed at network end points, for example, in applications.

The IAB, in 1996, summarized Internet architectural principles with three general philosophies: the objective of the Internet is global connectivity; the means for network level connectivity is the Internet Protocol; and intelligent functions should reside at end points rather than within networks.[37] This design philosophy was a departure from prevailing network approaches which established temporary fixed paths, or virtual circuits, between end points that remained fixed for the duration of a transmission. Part of the rationale for the end-to-end design was to allow applications to continue working in the event of a partial network failure.

Acknowledging that "Internet standards have increasingly become an arena for conflict," many IAB members expressed reservations about translation intermediaries like NAT.[38] Intermediary devices reduced the need for a single network protocol, IP, and would "dilute its significance as the single necessary feature of all communications sessions. Instead of concentrating diversity and function at the end systems, they spread diversity and function throughout the network."[39] The standards community feared that translation techniques would challenge older, dominant protocols and would create too many protocol choices for users. Interestingly, the original rationale for the end-to-end philosophy had included concern about "preserving the properties of user choice."[40]

36. Brian Carpenter, ed., "Architectural Principles of the Internet," RFC 1958, June 1996.
37. Ibid.
38. James Kempf and Rob Austein, "The Rise of the Middle and the Future of End-to-End: Reflections on the Evolution of the Internet Architecture," RFC 3724, March 2004.
39. Brian Carpenter, "Middleboxes: Taxonomy and Issue," RFC 3234, February 2002.
40. James Kempf and Rob Austein, eds., "The Rise of the Middle and the Future of End-to-End: Reflections on the Evolution of the Internet Architecture," RFC, 3724, March 2004.

Many IETF participants had become involved in Internet design when the network connected a relatively small group of individuals. This environment was quite different from later Internet contexts with widespread public and global access and the accompanying security challenges. Institutional and individual Internet users, in practice, began routinely implementing intelligent intermediaries that violated the end-to-end architectural principle. By 2000, network intermediaries, or "middleboxes," like security firewalls and translation devices became fairly widespread among US businesses and individual Internet users.[41]

Some IETF participants argued that the interruption of protocol formats by translation devices and other intermediaries would actually reduce the ability of users to implement security techniques, like encryption, which are specifically applied at end points. Others viewed NAT as the obvious remedy for Internet address exhaustion and a potential workaround for forestalling the transition to IPv6. Still others, including those involved in Asian and European IPv6 policies, ignored the prospect of address translation as an interim address conservation approach, instead leapfrogging to IPv6. Within the Internet standards-setting community, as Microsoft's Tony Hain described in 2000, NAT discussions "frequently take on religious tones," with proponents arguing NAT staves off IPv4 address depletion and dissenters referring to it as "a malicious technology, a weed which is destined to choke out continued Internet development."[42] The phenomenon of standards as a site of conflict, as Internet engineers themselves acknowledged, is certainly supported in the history of network address translation. Some in the standards community viewed IPv6 as a solution for minimizing network intermediaries that disrupted the end-to-end architectural principle.

## Internationalizing Internet Addresses

Another problem that Internet governance institutions sought to solve was the asymmetrical geographical distribution of addresses as historically unfolded in early IPv4 address assignments. The transition to a more distributed Internet registry system (though still under IANA with overall centralized address delegation responsibility) originated with the IAB in 1990.[43] This concern about greater internationalization emerged years

41. Tony Hain, "Architectural Implications of NAT," RFC 2993, November 2000.
42. Ibid.
43. Vinton Cerf, "IAB Recommended Policy on Distributing Internet Identifier Assignment and IAB Recommended Policy Change to Internet 'Connected' Status," RFC 1174, August 1990.

before the formation of ICANN. The initial IAB recommendation for a more international distribution of assignment functions arose from several circumstances—a concern about equitable regional distribution of addresses, an ever growing volume of assignments, the prevailing circumstance of the US government funding administrative activities supporting non–US entities, and, as addressed in chapter 2, a concern for retaining architectural control of the Internet by maintaining IP and IP addresses (vs. resources defined by other standards) as a unifying architecture.

IANA delegated some IPv4 addresses to internationally distributed regional Internet registries (RIRs) such as Asia's newly formed Asia Pacific Network Information Centre (APNIC) and Europe's Réseaux IP Européens–Network Coordination Centre (RIPE NCC). RIPE NCC was the first international registry. Headquartered in Amsterdam, RIPE NCC became fully operational in 1992. The Asia Pacific Network Interface Centre (APNIC), based originally in Tokyo but later relocated to Brisbane, Australia, assumed responsibility for allocating addresses to approximately 50 nations in the Asia Pacific region including Japan, China, Indonesia, and Australia. As the assignment function shifted to globally distributed registries, assignment in North America moved to a membership-oriented, nonprofit corporation called ARIN (American Registry for Internet Numbers), the RIR for much of the western hemisphere. As shown in figure 5.5, with the advent of the international registry system, the centralized IANA allocated addresses to RIRs, which in turn would reallocate address space to local Internet registries (LIRs), national Internet registries (NIRs), ISPs, or end user institutions.

According to an IPv4 address space audit the RIRs jointly conducted in 2002, APNIC controlled nine "/8" address blocks (IP addresses with a fixed 8-bit prefix; providing 16,777,216 addresses). Ignoring that APNIC allocated some of these addresses for exchange points and for experimental uses, the total allocated number of IPv4 addresses for all of the Asia Pacific region in 2002 totaled approximately 151 million, or roughly 3.5 percent of the IPv4 address space. China received a portion of this approximately 3.5 percent of IPv4 address space allocated to APNIC, as well as some other address allocations. Rather than operating local Internet registries (LIRs), China operated, beginning in 1997, a state registry called China Internet Network Information Center (CNNIC), run by the Ministry of Information Industry and operated by the Chinese Academy of Sciences (CAS). From a statistical perspective, the entire Asia Pacific region controlled a number of IP addresses roughly equal to one-tenth of the population of China, foreshadowing impending constraints.

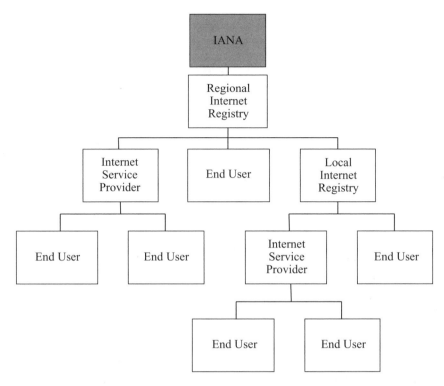

**Figure 5.5**
Address allocations and assignment structure

IANA, eventually a function under ICANN, still retained centralized coordination of the address space. ICANN formed in 1998 as a private, nonprofit corporation to administer the Internet's names and addresses and manage the Internet root servers. ICANN, consistent with Jon Postel's original responsibilities, would provide the following functions:

1. set policy for and direct allocation of IP number blocks to regional Internet number registries;

2. oversee operation of the authoritative Internet root server system;

3. oversee policy for determining when new TLDs are added to the root system; and

4. coordinate Internet technical parameter assignment to maintain universal connectivity.[44]

44. United States Department of Commerce, National Telecommunications and Information Agency, *Management of Internet Names and Addresses*, June 5,

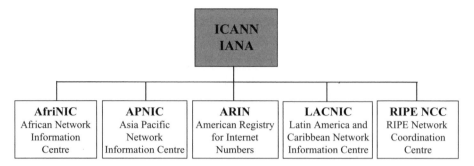

**Figure 5.6**
Regional Internet registries

Jon Postel and his close associates played a prominent role in the formation of ICANN, which subsumed the existing IANA. ICANN would ultimately have jurisdiction over the Internet address space. But the development of the RIR system sought to distribute control internationally and provide greater geographical equality in address distribution.

The RIRs are private, nonprofit institutions that employ a contract-oriented administrative model of governance. They serve large geographical areas, managing the address space allocated to them by IANA, under ICANN, and assigning addresses within their jurisdictional regions. Recall that in registry parlance, to allocate means to disperse addresses for subsequent distribution; to assign means to delegate addresses to ISPs and/or end users for actual use.

As figure 5.6 shows, two additional registries joined ARIN, RIPE NCC, and APNIC. ICANN formally recognized The Latin America and Caribbean Network Information Centre (LACNIC) as the fourth regional internet registry in October 2002.[45] The ICANN board formally accredited a fifth RIR, the African Network Information Centre (AfriNIC) in 2005 to distribute addresses within the African and Indian Ocean regions.

These five RIRs subsequently developed joint address registry policies establishing procedures for IPv6 address assignment.[46] The RIRs' joint

1998. Accessed at http://www.ntia.doc.gov/ntiahome/domainname/6_5_98dns.htm.

45. "Final Approval of LACNIC" in the Preliminary Report of the ICANN Board of Directors Meeting in Shanghai, October 21, 2002. Accessed at http://www.icann.org/minutes/prelim-report-31oct02.htm#FinalApprovalofLACNIC.

46. APNIC, ARIN, and RIPE NCC, "IPv6 Address Allocation and Assignment Policy," Document ID: RIPE-267, January 22, 2003.

registry procedures established "conservation" as one policy objective, calling for the elimination of wasteful practices and address stockpiling and requiring appropriate documentation to support all address requests.

One RIR principle directly departing somewhat from earlier IPv4 practices can be summarized in the following statement: "Address space not to be considered property."[47] Historically, once an organization received IPv4 address assignments, those addresses remained, in practice, an irrevocable possession of that organization, even if unused. To avoid the possibility of hoarding unused addresses, the RIRs agreed that it "is not in the interests of the Internet community as a whole for address space to be considered freehold property."[48] IPv6 addresses would be licensed rather than owned. RIRs would periodically renew these address licenses and retain the right to revoke addresses. This policy originated in the mid-1990s with the Internet Architecture Board and the Internet Engineering Steering Group, which issued recommendations for IANA about managing IPv6 address allocations.[49] The IAB/IESG position emphasized that a central authority (IANA) responsible for allocations was a necessary precursor of "good management" of the IPv6 address space. Additionally, allocations of address space by the IANA were not irrevocable, and there should continue to be no charges for addresses beyond fees to cover the administrative costs.

These administrative policies served not only to address the management of the IPv6 address space but to fortify the authority and philosophies of the prevailing Internet governance structure. First, IANA would retain exclusive centralized control of the address space, by delegation to registries. Second, even after delegating addresses to registries, IANA retained control because it could revoke allocations if, in its own judgment, it believed that an entity had "seriously mishandled the address space delegated to it."[50] The IAB also renewed its commitment to treating Internet address as a common public good by fortifying a system whereby IP addresses could not be bought and sold in open markets. Everyone would have a chance for Internet resources, not just the highest bidder. The belief that IP addresses were common pool resources in the public domain served as a philosophical underpinning for positions against

47. Ibid., sec. 4.1.
48. Ibid.
49. See IAB and IESG, "IPv6 Address Allocation Management," RFC 1881, December 1995.
50. Ibid.

exchanging IP addresses in open markets, a debate that would later emerge. Many in the standards and registry communities believed that "you cannot sell what you do not own."[51] This position preserved the power of the registries and of the centralized IANA to control the allocation and assignment of IP addresses rather than relinquish them to free markets.

On the other hand, IP addresses had ceased being completely free resources back in the mid-1990s. When ARIN was formally decoupled from the government-funded InterNIC in late 1997, it announced that it would commence charging for IP addresses, though only enough to cover the costs of its small assignment operation located in Chantilly, Virginia. ISPs accounted for a great number of IP address requests made to registries, and ARIN announced that new IP address requests would cost between $2,500 and $20,000 per year depending on the allocation size. The registry would not charge institutions holding existing IP addresses. Corporations (or individuals) requesting new IP addresses would pay a onetime fee depending on assignment size.

RIR policies, historically, have consistently and adamantly affirmed that they do not charge for IP addresses: "IP addresses are a shared public resource and are not for sale."[52] But, to cover operational expenses, the RIRs have charged initial allocation fees and maintenance fees for IP address allocations and assignments. The IP address fees have not varied significantly by RIR. For illustrative purposes, the following discussion uses a snapshot of LACNIC's pricing structure to describe the initial allocation cost and the annual renewal fees for ISPs to hold various size blocks of IP addresses. Recall that after the IETF developed classless interdomain routing (CIDR), address blocks were no longer allocated in Class A, B, and C blocks but in more flexibly sized network address increments. In post-CIDR terminology, a "/20" (pronounced "slash twenty") referred to an IPv4 address block with a 20-bit network number followed by 12 bits of host numbers, or a total number of IP addresses of $2^{12}$, or 4,096 addresses. A "/16" referred to an address block with a 16-bit network number followed by 16 bits of host address numbers, or 65,536 addresses. Table 5.3

51. Quote from ARIN Counsel Dennis Molloy documented in the minutes from the ARIN Members Meeting, October 16, 1998, section "Solicitations for the Purchase of Address Space." Accessed at http://www.arin.net/meetings/minutes/ARIN_II/index.html.

52. See RIPE NCC allocation and assignment policies available on the RIR's website. Accessed at http://www.ripencc.net/info/faq/rs/general.html#1.

**Table 5.3**
Sample IPv4 address registration prices

| | LACNIC IPv4 registration price list | | |
|---|---|---|---|
| Category | Size | Initial amount US$ | Renewal amount US$ |
| Small/micro | < /20 | $1,000 | $1,000 |
| Small | >= /20 y <= /19 | $2,000 | $2,000 |
| Medium | > /19 y <= /16 | $5,000 | $5,000 |
| Large | > /16 y <= /14 | $10,500 | $10,500 |
| Extra large | > /14 y <= /11 | $22,000 | $22,000 |
| Major | > /11 | $33,000 | $33,000 |

describes the pricing structure of one RIR—LACNIC—as a sample of IP address charges.[53]

The RIRs charged these IP address registration fees to large ISPs and LIRs that would, in turn, assign addresses to end users. The cost for end users directly purchasing from RIRs (some offer end user assignments) was considerably less than prices charged to ISPs. For example, LACNIC charged an annual maintenance fee of $400 to end users. ARIN charged the same initial registration fee for ISPs and end users but would not charge the large annual maintenance fee to users.

An interesting RIR fee schedule differentiation also emerged between IPv4 addresses and IPv6 addresses. In 2004 AfriNIC announced that "to encourage and promote IPv6 usage and allocation in the region, organizations which qualify to receive IPv6 allocation will have the first year's fees waived."[54]

The ARIN Board of Trustees similarly announced that it would waive IPv6 fees between January 1, 2005, and December 31, 2006. The RIR's IPv6 address policies sought both to promote IPv6 and to maintain the long-term viability of the IPv6 address space through conservation strategies. Some of these conservation policies underscored the ongoing power these entities, as well as lower level registries like LIRs and NIRs would have over addresses, and raised some potential concerns. One concern was the possibility of address reclamation abuse such as a national registry, closely aligned with a national government, reclaiming (i.e., seizing) an organization's addresses to retaliate for statements critical of the government. Similarly, a user organization requesting addresses from a local Internet registry

53. This chart reflects a snapshot of LACNIC's fee schedule in 2006.
54. Adiel Akplogan, "AfriNIC Fees Schedule (2004–2005)," May 10, 2004. Accessed at http://www.afrinic.net/docs/billing/afadm-fee200405.htm.

must provide justification for the request. The generality of such a policy leaves the door open for denials of address requests for almost any reason. Finally, the complete rejection of the prospect of exchanging some addresses in free markets (although charging nominal fees for addresses) seemed to eliminate the possibility of even opening up a dialogue about whether this type of exchange might serve to promote conservation rather than diminish conservation as the RIRs argued. But the possibility of IP address markets would gain increasing interest a few years later, in the face of looming IPv4 address space depletion, as will be discussed later in this chapter.

## Global Conflict over Internet Resources

The question of who should have the ultimate authority to distribute the global Internet resources of IP addresses has been a major source of tension and a central question in Internet governance. Not everyone was satisfied with the RIR system and the role of ICANN/IANA in allocating resources to RIRs for further distribution. In October 2004 the director of the ITU-T, Houlin Zhao, formally suggested a change in IPv6 address assignment procedures. Rather than RIRs acting as regional monopolies distributing addresses, Zhao proposed that blocks of IPv6 addresses be allocated to individual countries. Then governments would choose how to distribute addresses.[55] Entities seeking addresses could approach either the RIR or the government, producing some competition and choice in the IP address allocation system.

The ITU was not proposing that ICANN/IANA directly allocate IPv6 addresses to nations. Instead, the ITU would allocate blocks of IPv6 addresses to nations, giving the ITU significant IP address responsibilities. The ITU stressed its "unique position as an intergovernmental organization" under the United Nations[56] and the need for a legitimate governance organization responsible for resources and for establishing public policy. The ITU had traditionally established telecommunications standards and had handled such issues as radio spectrum disputes. In making his case for ITU influence on Internet governance issues, Zhao described the Internet as part of a broader existing public telecommunications

---

55. Houlin Zhao, "ITU and Internet Governance," draft input to the 7th meeting of the ITU Council Working Group on WSIS, December 2004. Accessed at http://www.itu.int/ITU-T/tsb-director/itut-wsis/files/zhao-netgov01.pdf.
56. Ibid.

infrastructure he called the "Next Generation Network (NGN)."[57] This subsumption of the Internet under a broader telecommunications infrastructure, rather than the inverse, would serve to bring Internet governance issues closer to ITU jurisdiction, with a constitution that described its mission "to maintain and extend international cooperation among all its member states for the improvement and rational use of telecommunications of all kinds."[58]

The IETF had led the development of the core routing and transport protocols for the Internet, but Zhao wished to contest the notion that the ITU had historically minimal involvement in the development of Internet standards or in Internet governance and administration. Zhao argued: "Some think that the ITU has no role in Internet standardization. But this is not correct."[59] He added that the ITU had been a "major contributor" to the Internet and Internet standards, making references to the ITU's involvement in access standards such as ADSL (Asymmetric Digital Subscriber Line) and cable modems, and standards directly related to specific applications of Internet voice transmission such as VoIP. Zhao claimed: "ITU activities have directly or indirectly, supported the technical development of Internet from the very beginning."[60]

The ITU offered another rationale for its proposed Internet oversight role. The ITU-T's director argued that the ITU could uniquely protect and represent the interests of developing countries relative to Internet governance because the ITU had traditionally defended the interests of developing countries. Zhao ultimately argued that the Internet's national importance necessitated management in each country by its national government. By Zhao's reasoning, governments should play a role at the international level, so presumably setting up an argument for United Nations governance of the Internet.

A controversy over control of Internet addresses and, especially Internet names, erupted in the summer of 2005 when Koffi Annan, secretary-general of the United Nations, announced the findings of a UN subgroup proposing several Internet governance alternatives that would in effect place Internet governance responsibilities under the United Nations. The United Nations "Working Group on Internet Governance," or WGIG,

57. Ibid.
58. Mission statement from International Telecommunications Union website. Accessed at www.itu.int on November 17, 2005.
59. See Zhao (2004) above.
60. Ibid., sec. 3.3.

issued the recommendations. Koffi Annan had established the WGIG in response to recommendations he received from the December 2003 World Summit on the Information Society (WSIS).[61] The group's mission had been to define Internet governance, identify major policy areas, and issue recommendations for Internet governance responsibilities in these areas.

The WGIG included 40 participants representing governments, the private sector, and individuals from what the United Nations called "civil society." Many of these participants held high level government positions in technology policy, such as Saudi Arabia's Deputy Governor of Technical Affairs for the Communications and Information Technology Commission of Saudi Arabia and Cuba's Coordinator of the Commission of Electronic Commerce.[62] Participants represented many countries, including Brazil, Egypt, China, Iran, Pakistan, Japan, Russia, Belgium, and others. Secretary-General Annan had the final authority in selecting the forty WGIG participants. The United States chose not to send a government representative to the WGIG.[63] Governments with patently restrictive Internet policies (e.g., Iran, Cuba, and Saudi Arabia) were prominently represented in this working group. Other participants were affiliated with a variety of commercial entities, academe, ICANN, the World Bank, and the ITU. No WGIG participants represented the US government, any US corporation, any organization involved in establishing standards for the Internet's routing and addressing protocols or DNS, or any leading private sector vendors (American or otherwise) involved in developing the products which incorporate Internet standards and policies. In other words, the UN group appeared to not incorporate the input of Internet users, Internet companies, or anyone technically involved in developing the systems underlying the policy areas the group addressed.

One of the charges of the WGIG was to define "Internet governance." After a lengthy exercise the group settled on the following definition:

61. The first phase of the World Summit on the Information Society was held in Geneva, Switzerland, on December 10–12, 2003.
62. The complete list of participants appears in the Annex of the WGIG's Report of the Working Group on Internet Governance, Chateau de Bossey, June 2005. Also see the United Nations Press Release, "United Nations Establishes Working Group on Internet Governance," PI/1620, November 11, 2004. Accessed at http://www.un.org/News/Press/docs/2004/pi1620.doc.htm.
63. Ambassador David Gross, US Coordinator for International Communications and Information Policy in the Bureau of Economic and Business Affairs, explained that the United States government did not participate in the WGIG because of "serious legal issues (under US law) that such participation could have raised," in a

Internet governance is the development and application by Governments, the private sector and civil society, in their respective roles, of shared principles, norms, rules, decision-making procedures, and programmes that shape the evolution and use of the Internet.[64]

On the surface the WGIG's definition of Internet governance might seem so broad as to be dismissed as a nondefinition. However, the definition conveyed some distinct Internet governance positions. The definition assigned an Internet governance role to "governments," setting up potentially greater involvement of national governments or intergovernmental entities such as the United Nations in regulating or administering Internet governance functions. Second, the definition assumed the existence of shared principles and norms in Internet policies. This assumption was not reflective of the political approaches to Internet governance among nations represented on the WGIG. The Internet governance principles and norms in Egypt, Cuba, China, Saudi Arabia, Pakistan, and Tunisia hardly resembled those of France, Brazil, and Switzerland in areas such as censorship, freedom of expression, privacy, surveillance, intellectual property, and Internet trade taxation. Finally, the WGIG definition of Internet governance itemized three entities—government, the private sector, and civil society—as responsible for Internet governance. The definition specifically did not single out technical and academic communities, historically influential in Internet governance roles such as standards setting. The presumed, tacit grouping of organizations such as the IETF in the broad "civil society" category, listed less prominently than governments, seemed to intimate a diminished role for technical communities.

The WGIG identified the following Internet governance policy issues: management of Internet resources (including IP addresses), network security, intellectual property and international trade, and Internet deployment in developing countries. Within these policy priorities, the highest priority for the WGIG was to address "unilateral control" by the US government in administering the root zone files of the domain name system. The WGIG also identified IP address allocation equitability by geographic area as a concern.

After developing its definition of Internet governance and identifying some specific Internet governance policy areas, the WGIG attempted to

---

State Department live Internet chat answering questions about the forthcoming WSIS summit in Tunis, November 2, 2005.

64. Report of the Working Group on Internet Governance, Chateuau de Bossey, June 2005. Accessed at http://www.wgig.org/docs/WGIGREPORT.pdf.

address who should assume responsibility in various areas. The group concluded that there currently existed a "vacuum within the context of existing structures, since there is no global multistakeholder forum to address Internet-related public policy issues."[65] The group determined that, in the forum that would fill this vacuum, no single government would have the ability to unilaterally act. As an interesting aside, this UN working group's emphasis on diminishing the dominance of the United States and eliminating unilateralism seemed, at the time, to mirror contemporaneous UN criticisms of what it described as US unilateral action in the US-led war in Iraq. The alternatives of multilateral Internet governance the WGIG explored involved, among other things, wresting the control of Internet addresses from the ICANN/IANA structures then overseen by the US Department of Commerce. The group also emphasized that "gender balance," or equal representation of men and women within any forum for discussions of Internet governance, should "be considered a fundamental principle." Some scholars and advocates noted that this recommendation lacked reflexive credibility considering the relatively few women within the WGIG discussing Internet governance as well as the limited rights of women in several WGIG countries.

The United Nations also alluded to a new approach for establishing Internet standards. The WGIG included standards development in a lengthy list of international government responsibilities.[66] The working group's recommendation seemed to be insinuating that Internet standards development move to an international, intergovernmental organization, presumably shifting standards development from the IETF to the UN-affiliated standards-setting body, the ITU. Furthermore, the recommended list of responsibilities for "civil society" and the private sector *did not* include standards development, excluding citizens, users, and vendors from governmentally constituted Internet standards development. Establishing top-down, intergovernmental, presumably UN-based control of Internet standards setting would represent a radical departure from the traditional Internet standards-development norms.

The WGIG also recommended four alternative models for multilateral Internet policy oversight. The first model would establish a Global Internet Council, anchored in the United Nations and comprised of governmental representatives to establish names and address policies such as how to

---

65. Ibid., sec. V.A.1.40.

66. Report of the Working Group on Internet Governance, Chateuau de Bossey, June 2005, sec. V.A.1.40. Accessed at http://www.wgig.org/docs/WGIGREPORT.pdf.

internationally allocate IPv6 addresses. Some of the recommendations included the following: completely eliminate the authority of the US Commerce Department in Internet oversight of the technical and operational functions of the Internet such as management of Internet addresses and the domain name system; either place ICANN under the United Nations or replace ICANN's role with a reformed internationalized organization, possibly given the name WICANN, (pronounced Y-CAN) for World Internet Corporation for Assigned Names and Numbers; and anchor any overarching international Internet governance council or forum in the United Nations.

The primary underpinning of these recommendations was to replace US oversight with UN oversight. The recommendations also raised questions about what role the private sector, Internet users, and Internet developers would have if a UN council led by governmental representatives assumed a greater role in Internet policy decisions. Another question was the possible architectural ramifications to the Internet if technical standards oversight related to addressing, routing, and the DNS moved from those historically involved in technical specifications to an intergovernmental organization. Finally, a major question was what impact greater Internet governance involvement of countries with repressive Internet policies would have on the Internet and access to knowledge generally. The Number Resource Organization[67] (NRO), a collaborative venture of the RIRs, acknowledged that the emphasis of the United Nations on multistakeholder models was important but suggested that the WGIG did not adequately present alternatives for existing organizations (e.g., the registries the NRO represents) to incorporate multistakeholder principles.[68] The NRO also accentuated the importance of retaining a role for academic and technical communities in Internet governance. The organization agreed that US oversight of ICANN and its IANA function must end but cautioned that any increase in government oversight might stunt innovation and increase bureaucracy.

A dominant and recurrent theme underlying the UN-proposed appropriation of Internet governance functions, including IPv6 address administration, involved the need for the Internet in the developing world. The

67. The Regional Internet Registries founded the Number Resource Organization (NRO) on October 24, 2003. The four RIRs extant at that time included: APNIC, ARIN, LACNIC, and RIPE NCC.
68. Number Resource Organization Document NR026, "Number Resource Organization (NRO) Comments on the WGIG Report," July 2005. Accessed at http://www.nro.net/documents/nro26.html.

articulated rationales by the United Nations for recommending a diminishment of US power did not specifically address economic and political requirements of the developed countries (represented on the WGIG) to gain more Internet governance responsibility such as oversight of critical Internet resources. Instead, the WGIG agreed upon two overarching requirements for Internet governance legitimacy, both related to developing countries: the "effective and meaningful participation of all stakeholders, especially from developing countries" and the "building of sufficient capacity in developing countries, in terms of knowledge and of human, financial and technical resources."[69]

The United Nations emphasized the priority of Internet capacity-building as a mechanism for helping developing countries and as a rationale for more multilateral control of Internet governance, including management of the IP address space. Arturo Escobar stresses that "understanding the discursive and institutional construction of client categories requires that attention be shifted to the institutional apparatus that is doing the 'developing.'"[70] The WGIG and the United Nations portrayed developing countries as targets for intervention. These institutions also prescribed themselves as solutions to these problems. For example, the United Nations framed the appropriation of Internet governance functions from the United States as a necessary precursor to legitimate developing country representation and resource distribution.

At the time of WGIG's proposals and accompanying rationale that a more equitable resource and governance structure was necessary for developing countries, what was the status of the global distribution of IPv4 and IPv6 addresses? By the summer of 2005, address allocation statistics appeared geographically more egalitarian than in prior years. IPv4 addresses were geographically distributed equally among the Asia Pacific region, North America, and Europe, with small allocations to Latin America and Africa. Europe and the Asia Pacific region controlled the majority of IPv6 address allocations. Figure 5.7 illustrates the IPv4 and IPv6 address allocation statistics from 2005. According to the address distribution statistics, Africa and Latin America controlled only 4 percent of IPv4 addresses and 4 percent of IPv6 addresses.

Two weeks before the United Nations released its Internet governance report advocating US relinquishment of unilateral oversight of Internet names and addresses, the US Commerce Department, on behalf of the Bush

69. Report of the Working Group on Internet Governance.
70. Arturo Escobar, *Encountering Development, The Making and Unmaking of the Third World*, Princeton: Princeton University Press, 1995, p. 107.

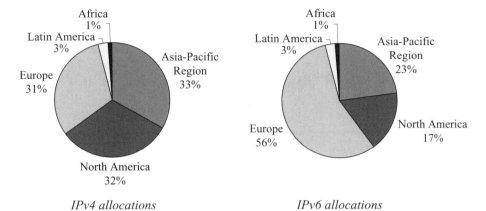

*IPv4 allocations*                          *IPv6 allocations*

**Figure 5.7**
Regional IP address allocations

administration, issued an articulation of core principles for the Internet's addressing and domain name systems. The "US Principles on the Internet's Domain Name and Addressing System"[71] asserted that the US government would retain its historical responsibility and oversight of ICANN. Recall that ICANN's primary responsibilities included central administration of Internet addresses through its IANA function, the operation of the Internet's root name server system, and administering domain names. The primary message in the Commerce Department's articulation of Internet principles was that US unilateral oversight of addresses and DNS administration would continue, cutting off the possibility of internationalizing this function by relinquishing any responsibilities to the United Nations. The US argument for maintaining the status quo rested on the notion that the current Internet system was working and that any changes might disrupt the security, stability, and efficient operation of the Internet.

The Bush administration's statement of principles conveyed an impression of durability and firmness because it would serve as a guiding foundation for establishing all federal government policies related to Internet names and addresses "in the coming years."[72] The new principles also emerged as one part of a broader administration technology framework.

71. "US Principles on the Internet's Domain Name and Addressing System." Accessed at http://www.ntia.doc.gov/ntiahome/domainname/USDNSprinciples_06302005.htm on December 8, 2005.
72. According to the website of the National Telecommunications and Information Agency. Accessed at http://www.ntia.doc.gov/ntiahome/ntiageneral/bios/mdgbio.htm on December 4, 2005.

Assistant Secretary of Commerce Michael Gallagher directed the policy review effort leading to the formation of the US principles. President Bush had appointed Gallagher on July 1, 2004, to the post of assistant secretary of commerce for communications and information and administrator of the National Telecommunications and Information Administration (NTIA).

Assistant Secretary Gallagher announced the new US policy principles during his presentation at the Wireless Communications Association annual conference in Washington, DC, on June 30, 2005.[73] The presentation emphasized three areas critical to continued US economic success: broadband, spectrum policies, and the Internet. Gallagher linked the administration's policies of business tax relief and regulatory reductions with economic growth in broadband. He also identified spectrum management reform geared toward freeing up scarce resources of radio frequencies as a precursor to promoting the growth of wireless broadband technologies and increasing imports of these products to vast markets like China and India. Finally, Gallagher stated that the Department of Commerce would retain its role in Internet name and address system oversight to preserve the Internet's economic stability, economic opportunities, and security.

The Bush administration's position embraced the status quo, but it was also a reversal of previously established policy directives. Beginning with the Commerce Department's 1998 White Paper[74] calling for the creation of a private, nonprofit corporation to administer the Internet's domain name and addressing functions, US government policy included transition agreements with ICANN anticipating an eventual phasing out of a federal government role in Internet address and name system oversight. The plans for a transition from federal government control originated during the Clinton administration and had two primary objectives: a more privatized approach and more internationalized oversight. The US Department of Commerce had anticipated that US government policy oversight of the new private corporation would end within two years: "the US government would continue to participate in policy oversight until such time as the

73. The NTIA website published Assistant Secretary Gallagher's presentation. Accessed at http://www.ntia.doc.gov/ntiahome/speeches/2005/wca_06302005_files/ frame.htm on December 20, 2005.
74. United States Department of Commerce, National Telecommunications and Information Agency, Docket Number 980212036-8146-02, *Management of Internet Names and Addresses*, June 5, 1998.

new corporation was established and stable, phasing out as soon as possible, but in no event later than September 30, 2000."[75]

The Commerce Department's original policy objective established that the functions related to administering the names and number systems would be private, nonprofit, and "managed by a globally and functionally representative Board of Directors."[76]

The policy anticipating a phasing out of federal government oversight required ICANN to meet certain conditions and went through several years of evaluations followed by extensions of federal government oversight. For example, in 2003 the policy agreements between the US Department of Commerce and ICANN anticipated an eventual phasing out (by 2006) of US governmental funding and oversight of the new entity.[77] The new Commerce Department declaration of Internet principles reversed this. Against the backdrop of the UN proposing an eradication of unilateral US Department of Commerce oversight, the United States formally reversed its transition objective and drew a demarcation preserving its oversight role indefinitely.

In addition to preserving boundaries, the US Declaration of Principles appeared to also anticipate and rebuff the possibility of the United Nations assuming any Internet governance role. The statement of principles stated that no single organization could adequately "address the subject in its entirety." The notion of a variety of organizations rather than a single forum as appropriate for Internet governance preempted the UN impending report seeking Internet governance power. Finally, the US principles appeared to prioritize the possible role of market-based approaches and the private sector, promising "the United States will continue to support market-based approaches and private sector leadership in Internet development broadly." Market-based approaches were not historically pertinent to the Internet names and numbers management function, but this principle served to diminish the prospect for greater governmental (or intergovernmental) involvement while maintaining overall US oversight of Internet resources.

The timing of the announcement preceded the WGIG report by two weeks and also ensconced a firmer position from which the United States could

75. Ibid.
76. Ibid.
77. See the "Memorandum of Understanding between the US Department of Commerce and the Internet Corporation for Assigned Names and Numbers, Amendment 6," September 2003, http://www.ntia.doc.gov/ntiahome/domainname/agreements/amendment6_09162003.htm.

negotiate during an upcoming UN-sponsored conference discussing Internet governance issues. The Internet Governance Project (IGP), a consortium of prominent Internet governance scholars from Syracuse University, Georgia Institute of Technology, and elsewhere criticized the Bush administration's announcement. The IGP called Assistant Secretary Gallagher a newcomer to the debate who didn't realize that what he called the US government's historic involvement was less than seven years old. It is true that ICANN was only seven years old, but the US government had historically maintained some oversight and funding of the responsibilities it later repositioned under ICANN. Nevertheless, the IGP's position suggested that oversight, albeit limited oversight, of the ICANN functions must be internationalized and that "No single government can be trusted to eliminate all considerations of national self-interest from its oversight role."[78]

ICANN's legitimacy emanated from increasing international representation and the expectation that US unilateral oversight would eventually wane. A continuation of US unilateralism might detract from ICANN's already tenuous legitimacy and create conditions whereby the Internet might fragment into national segments independent of US participation. In short, "If nothing changes, the US role will continue to inflame political criticism of Internet governance for years to come."[79] The US announcement did appear to incite political criticism. In one graphic example, British technology weekly *The Register* framed the US announcement in an overall cultural context of the Bush administration's world philosophy: "that the US will continue to run the Internet and everyone will just have to lump it—is very in keeping with how the US government is currently run!"[80]

Once Koffi Annan formally released the WGIG report, the US Department of State released official "Comments of the United States of America on Internet Governance"[81] responding to the findings and recommendations. Without specifically stating that the Commerce Department

78. Internet Governance Project Concept Paper, "The Future U.S. Role in Internet Governance: 7 Points in Response to the U.S. Commerce Dept.'s 'Statement of Principles,'" July 28, 2005. Accessed at http://www.internetgovernance.org/pdf/igp-usrole.pdf.

79. Ibid.

80. Kieren McCarthy, "Bush Administration Annexes Internet," *The Register*, July 1, 2005.

81. US Department of State, Bureau of Economic and Business Affairs, "Comments of the United States of America on Internet Governance," August 15, 2005. Accessed at http://www.state.gov/e/eb/rls/othr/2005/51063.htm on November 11, 2005.

planned to retain its ICANN oversight role, the State Department echoed the sentiments expressed in the US principles on Internet governance. The State Department suggested an implausibility of one single entity completely addressing the spectrum of Internet governance issues and included references to various global Internet governance entities (e.g., the World Intellectual Property Organization, or WIPO, and the London Action Plan on spam) into its response. The State Department also disputed the notion that Internet governance related to address and name administration was completely centralized or unilaterally administered. Internationalization and administrative distribution of the Internet was evident in the creation of RIRs, the efforts to allocate IP addresses in a more geographically equitable pattern, and because the "vast majority" of the 103 root servers (and mirror root servers) were located outside of the United States. The State Department's formal comments were diplomatically phrased in not specifically denouncing (or even mentioning) the possibility of UN Internet oversight but nevertheless presented arguments that would countervail any potential governance change. For example, the document reiterated US commitment to freedom of expression, presumably an argument against direct Internet governance participation by countries such China and Cuba through UN conduits. The State Department also acknowledged the need for governmental representation but highlighted the importance of civil sector and private sector involvement in Internet governance, using as an example the private sector led ICANN with government input provided through ICANN's Global Advisory Committee (GAC) in contrast to UN oversight, which could limit civic involvement and could impede private investment, competition, and associated innovation.

## International Impasse

The United Nations and the United States espoused seemingly irreconcilable differences about Internet governance, including, among many functions, the IP address oversight role. The United States declared it would continue its ICANN oversight function and the United Nations declared US unilateral oversight must cease. The international debate over which entity should oversee Internet addresses and the domain name system continued in "PrepCom3," the third preparatory committee meeting prior to the World Summit on the Information Society (WSIS) scheduled for November 16, 2005, in Tunis, Tunisia. PrepCom3, held in September in Geneva, Switzerland, was a politically charged, two-week session of debates about Internet governance and other Internet issues.[82] Nearly 2,000 individuals representing governments, nongovernmental organizations, and businesses participated in the sessions,[83] including a US delegation with David Gross, US coordinator of International Communications and Information Policy in the Department of State. The preparatory conference ended with a polarizing impasse over the Internet governance issue of management of Internet addresses and the domain name system, reflecting prevailing tensions between US and UN positions.

The US and UN positions shared one common denominator in invoking democratic ideals as justifications for each argument: a linkage between Internet architectural oversight and democratic freedoms. This linkage resembled prevailing associations, among IPv6 advocates, between the IPv6 standard and the promotion of worldwide freedom and democracy throughout the world. Multilateral oversight by a UN-based entity was the true democratic approach, according to those espousing the diminishment of US oversight. Others argued that handing over Internet oversight to an organization—the United Nations—with no democratic preconditions for membership could compromise the democratic and libertarian underpinnings of the Internet.

Some in the US Congress supported the Bush administration's position on Internet governance by formally denouncing the prospect of UN intervention. Senator Norm Coleman (R-MN) entered a statement into the Congressional Record censuring the recommendation in the WGIG report calling for an end to US oversight of ICANN functions. Coleman, with Senator Dick Lugar (R-IN), had recently introduced UN reform legislation, the Coleman–Lugar UN Reform Bill, which addressed a "culture of corruption" at the United Nations centered around the Oil for Food scandal. Coleman described UN management as "at best, incompetent, and at worst corrupt" and denounced the possibility of UN control over the Internet.[84] Besides the negative heuristics of mismanagement and corruption, Senator Coleman argued that the move would allow countries like China and Cuba, with no commitments to democratic freedoms or the free flow of information, to gain unwarranted influence over the Internet.

82. The ITU provided video web casts of PrepCom-3 on its website. Accessed at http://www.itu.int/wsis/preparatory2/pc3/#pc3.
83. According to the final list of participants, PrepCom-3 for World Summit on the Information Society, Geneva, Switzerland, September 2005. Accessed at http://www.itu.int/wsis/docs2/pc3/participants-list-final.pdf.
84. From "Coleman Denounces Report Calling for UN Global Internet Control: Coleman Opposed to Any Proposal to Hand Control of Internet Governance over

Three members of the House of Representatives, California Republican John Doolittle, Virginia Republican Bob Goodlatte, and Virginia Democrat Rick Boucher, issued a similar resolution offering more political backing for the administration's position opposing involvement of the United Nations in ICANN oversight. The House resolution concurred with previously issued US principles on Internet governance and stated that any interest in moving the name and addressing system under UN control was "on political grounds unrelated to any technical need."[85] Additionally, the resolution argued that US oversight of names and numbers should continue for the following reasons: historical roots of the Internet in US government funding, retention of private sector leadership and public involvement as essential for continued Internet evolution, maintenance of the Internet's security and stability, and preservation of freedom of expression and free flow of information.[86] The general political position of the Bush administration and some in Congress argued that ICANN, while imperfect, allowed for significant private sector involvement and that international representation and any transfer of ICANN functions to the United Nations would threaten democratic freedoms of the Internet, private sector involvement, and the stable ongoing operations of the infrastructure.

After lengthy preparatory meetings, working group deliberations, and great controversy, the aftermath of the ITU-organized World Summit on the Information Society (November 2005) as it pertained to address oversight, included retention of the status quo. The summit's consensus statement, "the Tunis Agenda for the Information Society"[87] made no specific mention of ICANN or the United States but preserved the status quo by leaving Internet resource control in the existing governance forums, meaning ICANN with US government oversight. The summit rejected the WGIG recommendation to create a new UN-based governance body, primarily because changes could not proceed without the agreement of the United States, which would not acquiesce to any structural changes. On the final day of the summit, John Marburger, presidential science and technology advisor, firmly reiterated the US position to retain the existing oversight structure. The US State

to the United Nations," published on Senator Coleman's website on July 29, 2005. Accessed at http://coleman.senate.gov/ on December 2, 2005.

85. HCON 268 IH, 109th Congress, House Congressional Resolution 268, "Expressing the Sense of the Congress Regarding Oversight of the Internet Corporàtion for Assigned Names and Numbers," October 18, 2005.

86. Ibid.

87. Final WSIS documents, conference statements, and videocasts published on the ITU website. Accessed at http://www.itu.int/wsis.

Department described the rejection of a new UN-based governance body as a victory that would "keep the Internet free of bureaucracy."[88] Not surprisingly, ICANN welcomed the WSIS Tunis Declaration, and suggested the WSIS recognition of ICANN's multistakeholder model (i.e., its Governmental Advisory Committee) would ensure the ongoing stability and integrity of the Internet's name and addressing system. The WSIS statement included a compromise that many nations described as a victory for multilateralism, the formation of an Internet Governance Forum (IGF). The IGF would continue the dialogue about Internet governance issues but would have no decision-making authority. The major concerns and issues raised at WSIS remained open and unresolved.

The question of management of critical Internet resources was a central discussion topic at the inaugural IGF held in Athens, Greece, in 2006; the second IGF in Rio de Janeiro, Brazil, in 2007; and the third IGF in Hyderabad, India, in 2008. The purpose of the IGF was to create a formal space for multistakeholder policy dialogue to address issues related to Internet governance, facilitate discourse among international Internet governance institutions, and to promote the engagement of stakeholders, particularly developing countries, in Internet governance mechanisms. The topic of "Critical Internet Resources" has represented one prominent IGF track. Discussions about critical Internet resources have focused on three areas: the root server system, domain names, and Internet addresses. The most pervasive questions about critical Internet resources revolved around questions of who should control them, particularly the role of the US government in ICANN oversight, the potential role of national governments in further entering or exiting ICANN's decision-making structure, and finally the potential role of the ITU, as part of the United Nations, in entering Internet governance of critical Internet resources. The multistakeholder dialogue seemed to do nothing to end the ongoing impasse over control of global Internet resources.

## Prospects for a Market Solution

The Internet standards community and the regional Internet registries have traditionally espoused the philosophy that Internet addresses are a shared public resource and that it would be inappropriate to create an

88. "World Summit Agrees on Status Quo for Internet Governance." Accessed on U.S. State Department website http://usinfo.state.gov/eur/Archive/2005/Nov/16-493027.html on December 12, 2005.

environment in which these resources are bought and sold in free markets. But by 2008 consensus projections about the impending depletion of the IPv4 address space, without some radical intervention, forecasted that addresses would be completely exhausted as early as 2011. This looming problem provided an opening for a reevaluation of the long-standing philosophy about IP addresses as shared public goods and generated new interest in the possibility of introducing Internet address markets in which addresses could be bought and sold. Many organizations have enormous blocks of IP addresses, and a large percentage of these addresses are unused but unavailable for others.

Creating Internet address markets, on one hand, has pragmatic appeal. Address exchanges would theoretically provide economic incentives for Internet address holders to free up unused addresses. Through some sort of a market exchange, possibly coordinated by RIRs and overseen by IANA, these addresses could be sold to those requiring addresses. This interjection of additional Internet addresses into markets could, in the very short term, stave off Internet address depletion. It would also potentially be a preemptive strike aimed at preventing black market trade of Internet addresses outside the purview of Internet registries, although some trade, or at least "gaming the system," in IP addresses already occurs.[89] Finally, address markets could further forestall the transition to IPv6 (or some other solution to address scarcity), providing additional time for technological development and innovations in transitional technologies designed to facilitate IPv4 and IPv6 interoperability.

There have been many arguments against the concept of Internet address markets. One concern is that it would hinder what many believe is the inevitable migration to IPv6. This argument has weaknesses. On the surface it makes sense that the longer IPv4 addresses are available, the less incentive there will be to deploy IPv6. But, as IPv4 addresses become scarcer, and allowed to be exchanged in address markets, the more expensive they will become. This greater cost would potentially provide incentives for markets to adopt IPv6 rather than IPv4. Rather than postponing IPv6 upgrades, this might provide greater incentives to deploy IPv6 rather than IPv4. Another concern about introducing address exchanges is that creating a global address market would require Internet

89. See Milton Mueller, "Scarcity in IP addresses: IPv4 Address Transfer Markets and the Regional Internet Address Registries," Draft Internet Governance Project White Paper, July 2008. Accessed at http://internetgovernance.org/pdf/IPAddress_TransferMarkets.pdf.

address registries to coordinate and agree upon the terms for these exchanges.

There are other concerns about the possible introduction of Internet address exchanges. First, it would be important to ensure that those buying Internet addresses would really use them or provision them to their customers rather than hoarding them or "flipping" IP addresses to a higher bidder. Exchanges could also increase global inequities in address distribution rather than repair them. Address markets would be advantageous to those who could afford to pay premiums for global Internet addresses. Those in emerging markets, already with a smaller share of Internet addresses, might not benefit. Those who could not afford to pay these premiums would be forced to, for economic reasons, "go it alone" as early adopters implementing IPv6-only solutions, with all the challenges associated with backward compatibility, complexity, and lack of applications. Another concern is that Internet address exchange markets might attract interest in direct governmental regulation of Internet addresses, a development that would change the character of Internet address management, face enormous transnational jurisdiction challenges, and have its own set of unintended consequences.

By 2008 three RIRs—APNIC, RIPE NCC, and ARIN—began considering proposals for address exchange markets. Even entertaining the possibility of this market-based approach represented a dramatic departure from earlier institutional positions with regard to Internet addresses as global public resources that should not be owned.

The issue of address markets is somewhat of a red herring because it is inherently only a very short-term amelioration of a long-term problem. Whether IP address exchanges are implemented or not does not significantly adjust predictions that the IPv4 address space will become critically scarce. The combination of new mobile and multimedia applications that require IP addresses coupled with the global growth of traditional Internet access will require a more long-term solution than address exchanges, whether massive deployment of translation or a longer term upgrade to IPv6.

### Thoughts about the Sufficiency of the IPv6 Address Space

Internet addresses provide an example of how technical standards create scarce resources and how, once the value of these resources is understood, they become a power struggle among those seeking to retain or obtain greater control and economic positioning relative to these resources.

Explaining the sudden value of electromagnetic spectrum during the nine-teenth-century expansion of radio technologies, economist Hugh Aitken said: "Here we have new resources—invisible resources, to be sure. . . . These resources furthermore, when their economic and military uses came to be appreciated, were to become the object of competitive struggles for exclusive possession and occupancy, just like the colonial empires carved out by European powers in North America in the seventeenth century or in Africa in the nineteenth."[90] Like radio spectrum, Internet addresses came to be seen as invisible, but valuable, scarce resources.

The original ARPANET destination codes were only 5 bits long, providing a total of 32 unique addresses. Researchers gradually augmented the number of addresses as they anticipated requirements for connecting addi-tional devices. IPv4 specified a 32-bit code providing more than four billion unique addresses. Original administrative and technical decisions such as the Internet class system, assignment inefficiencies, and an asymmetrical allocation to US institutions contributed, along with rapid global Internet growth, to concerns about an impending IPv4 address shortage. However, CIDR, NAT, address conservation policies, and the distribution of large blocks of the IPv4 address space to international registries helped mitigate some concerns about address depletion and inequity. IPv6 advocates, including governments in Asia and the European Union, have described IPv6 and the abundance of available IPv6 addresses throughout the globe as the solution to any conceivable address depletion concerns.

Analogous to the question of who would be responsible for Internet standards that had shaped the selection of SIPP over the ISO-based alterna-tive as the next generation Internet protocol, the issue of who would ultimately control IP addresses shaped decisions about the address assign-ment structures. Tensions between those involved in the Internet since the early days of ARPANET and newer participants, and politically reflective tensions between an American-controlled structure and greater multi-lateral control, have continued to fuel controversies about institutional administrative control. A single individual originally distributed addresses. As this responsibility shifted to more formal institutional structures, Postel and his colleagues remained central figures in structural decisions regard-ing resource distribution. The ongoing institutional decision to oppose the possibility of exchanging IP addresses in free markets served to support the technical community's philosophy that Internet resources be available to

90. Hugh G. J. Aitken, *Syntony and Spark: The Origins of Radio*, New York: Wiley, 1976, p. 32.

everyone but also fortified the centralized institutional control of resource distribution.

The history of the address space also provides clear lessons about the unintended consequences of technological development and about an almost universal inability to foresee new applications or predict how the Internet will expand or change over time. Historical predictions about when the IP address space would become completely depleted have consistently been incorrect. Recall that the initial prediction from some in the Internet's technical community cited 1994 as the target date for complete Internet address depletion. Technical changes such as NAT and CIDR contributed to the endurance of the IPv4 address space, as have address conservation policies.

Predictions about the imminent deployment of IPv6 have been as consistently incorrect as predictions about the depletion of the IPv4 address space. The sufficiency of the IPv6 address space to meet the Internet's future requirements is also an unquestioned assumption. Will this also be incorrect? Most concerns have centered on IPv4 addresses while the universal assumption about the IPv6 address space is that it is an almost boundless resource to accommodate future Internet growth into the foreseeable future. Expectations about the adequacy of the IPv6 address space mirror expectations about the adequacy of the IPv4 address space twenty years earlier. Billions of addresses appeared extravagant in the era in which the IPv4 standard emerged but, retrospectively, seems parsimonious because it provides less than one Internet address per human on earth. In contrast, the IPv6 standard, by specifying 128-bit addresses, theoretically provides $2^{128}$ unique addresses. One way to describe this number is with scientific notation: the standard allows for a theoretical maximum of $3.4 \times 10^{38}$ unique addresses. The multiplier undecillion can also help describe this number: the standard allows for a theoretical maximum of 340 undecillion addresses. In the American system an undecillion is mathematically equivalent to $10^{36}$. The British system of multipliers sometimes uses quintillion. Descriptions of the size of the IPv6 address space are inconsistent, ironically, because of the lack of universal standards for mathematical multiplier terminology. For example, a quintillion in the American system equals $10^{18}$. A quintillion in the British system equals $10^{30}$. Even discussions about IPv6 require translation.

What most cultures have agreed upon is an analogy to describe the size of the IPv6 address space. The number of IPv6 addresses is equal to the number of grains of sand—depending on the source—on our planet earth, on 300 million planets the size of the earth, or in the Sahara desert. For example, the European Commission's 2002 IPv6 strategy announcements included a reference to the size of the IPv6 address space as supporting, "more locations in cyberspace than there are grains of sand on the world's beaches."[91] Technology vendors, IPv6 Forums, and the technical media have consistently used this "grains of sand" analogy to describe the IPv6 address space. This analogy conveys the impression that IPv6 provides a boundless reserve of addresses to meet Internet requirements for the conceivable future. Interestingly, the Latin word for sand is *arena*, a locus of battle and competition. Given the history of the Internet, it is surprising that there is not more concern about the IPv6 address space becoming as contentious as the IPv4 address space. Circa 1981, no one envisioned a possible scarcity of IPv4 addresses. Two decades later, IPv6 proponents appear to not conceive of the possibility of future constraints on the IPv6 address space.

This assumption that the Internet will never face address constraints overlooks the history of the Internet itself. Scientist Leonard Kleinrock, one of the original ARPANET developers beginning in the late 1960s, has a long-term perspective on the evolution of increasing demands on the Internet address space. Twenty-five years after his initial ARPANET involvement, Kleinrock, in public remarks, raised questions about the adequacy of the IPv6 address reserve. Kleinrock asked, "Why does IPv6 only have 128 bits?" He suggested that, although it seemed adequate at the time, it might "run into trouble two decades from now."[92]

91. European Commission Press Release, "Commission Takes Step towards the Next Generation Internet," Reference IP/02/284, Brussels, Belgium, February 2002. Accessed at http://europa.eu.int on April 2, 2004.
92. Leonard Kleinrock, public remarks during final panel discussion at the United States IPv6 Summit, Arlington, VA, December 2004.

# 6   Opening Internet Governance

The principle of constant change is perhaps the only principle of the Internet that should survive indefinitely.[1]

—RFC 1958

One of the most significant developments in the history of the Internet has been the near exhaustion of the Internet address space. The successful growth of the Internet, more than anything else, has consumed many of these 4.3 billion Internet addresses. But institutional policies, technical design choices, and uneven geographical distribution of addresses have also played significant roles. Internet address scarcity was not a sudden phenomenon, but a gradual development foreseen by Internet designers beginning in 1990, long before widespread public Internet use. The Internet Protocol, with its 32-bit address length, created the Internet address space in 1981. The Internet standards community developed the next generation Internet protocol, IPv6, to exponentially expand the pool of Internet addresses. It has yet to be seen whether this protocol, available for more than a decade, will gain traction and whether its adoption will sufficiently solve the problem of Internet address exhaustion.

What is historically clear is that IPv6 adoption has not unfolded as its designers and advocates expected or as government mandates dictated. The success of IPv6 requires some degree of critical mass. Internet administration may be somewhat centralized, but Internet adoption is completely decentralized. The most economically powerful Internet users with the greatest ability to influence markets have always possessed sufficient addresses and have not had any incentive to upgrade to IPv6.

1. Brian Carpenter, ed., "Architectural Principles of the Internet," RFC 1958, June 1996.

Technological factors have also hindered the adoption of the new protocol. IPv6 is not directly backward compatible with IPv4. In a design context assuming that all users would upgrade to the new protocol for the common good, backward incompatibility would have seemed like less of a problem. In reality, users have resisted the resource-intensive move to IPv6 unless they expected some immediate benefit or payback.

Historian of technology Thomas Hughes has said that overcoming the conservative momentum of a large technological system requires a force analogous to that which extinguished the dinosaurs, such as the 1970s oil embargo or a catastrophe such as the Challenger space shuttle tragedy or the Three-Mile Island nuclear disaster.[2] The complete depletion of the Internet address space might provide the tipping point for more widespread IPv6 adoption. Alternatively, a new "killer application" that requires more Internet addresses could drive its adoption. Regardless, the translation approaches IPv6 was designed to displace will likely become much more prevalent. The slow deployment of IPv6, along with the impending depletion of the IPv4 address space and the increase in translation techniques, will have social and political implications as well as repercussions for the Internet's technical architecture and possibly even Internet governance structures.

Internet protocols and the resources they create are the least visible but arguably most critical component of the Internet's technical and legal architecture. The development of universal Internet protocols and the management of scarce resources are fundamental Internet governance responsibilities. These are not purely technological functions but activities with significant public policy implications. This examination of the "next generation Internet protocol" has described how the design, implementation, and use of Internet protocols can have significant economic and political implications, as well as technical.

IPv6, as the promised evolution of the Internet's key underlying protocol, has served as a locus for international tensions over globalization and control of the Internet. Internet designers selecting the new Internet protocol in the 1990s established a guideline that the evaluation would be based on technical requirements rather than influenced by institutional or political concerns. Yet the issue of what institution—the IETF or ISO— would control the standard was a factor in the selection process. In the

2. Thomas Hughes, *American Genesis: A History of the American Genius for Invention*, New York: Penguin Books, 1989, pp. 462–63.

context of Internet globalization, the decision reflected tensions between competing corporate interests as well as between a then-predominantly American institution and a more international standards institution. As the Internet's development environment transformed from a small community of trusted insiders to a more diffuse international collaboration, the selection of IPv6 solidified the authority of the IETF to continue managing core Internet standards. Centralized control of IP addresses similarly developed into a political impasse between retaining American oversight and pursuing greater multilateralism. IPv6 design choices faced by Internet engineers also made decisions about the public's civil liberties online, particularly issues of individual privacy on the Internet.

Protocol adoption strategies also extend far beyond technical considerations to reflect concerns about global economic competitiveness, national security, and even global democratic freedoms. The intellectual property arrangements underlying protocols can have significant global trade implications and determine how innovation and competition will proceed within technology markets. Furthermore, technical protocols, including the Internet Protocol, create the scarce resources necessary for equitable participation in the global information society.

These three spheres of protocol development, adoption, and technical resource distribution have a common denominator in that decisions made in these areas self-consciously occurred outside of classical market mechanisms. The Internet standards community selecting IPv6 overlooked the short-term views of some large corporate US Internet users, with enormous IPv4 installed bases, who were reluctant to upgrade to a new standard. When Internet users were mutually familiar with each other and were also Internet developers, user technical development and standards selection was the norm. Users were also standards developers. Users eventually became a more amorphous "market," severing the connection between users and standards development. The technical community believed that the decision was "too complicated for a rational market-led solution"[3] and that "we still need Computer Science PhDs to run our networks for a while longer."[4] User-developer Internet standards governance was acceptable

3. From the Minutes of the IPng Decision Process BOF (IPDecide) reported by Brian Carpenter (CERN) and Tim Dixon (RARE) with additional text from Phill Gross (ANS), July 1993.
4. Brian Carpenter, submission to big-internet mailing list, April 14, 1993.

when users were PhD computer scientists but not when users became a more generalized, corporate, and public market.

The historical distribution of IP addresses, which, on the surface, could be a straightforward problem of supply and demand of common pool resources, followed a similar trajectory. IP addresses have never been exchanged in free markets and were originally generously allocated, in enormous blocks, on a first come–first served basis to American organizations involved in early Internet development. Address distribution evolved over time to be more internationally equitable, but throughout its history, address distribution has never had a free market basis. Regarding IPv6 adoption, many governments issued top-down, national IPv6 mandates. With the exception of the initial American position to "let the market decide," state interventions rejected competitive market mechanisms in both development and deployment of IPv6 products, instead attempting to federalize technology selection for citizens and institutions. Bearing in mind that state procurement practices are a component of markets, these state IPv6 mandates seemed incongruous with broader market reluctance to embrace IPv6 as expected. It was also interesting how the standards community emphasized the philosophy of bottom-up, grassroots standards selection while governments issued top-down mandates, even while both approaches rejected free market mechanisms.

This chapter generalizes this discussion into a framework for understanding the political and economic implications of technical protocols. Given the social importance of technical standards, questions about who sets these standards and by what process are of great consequence. This chapter recommends a framework of best practices in Internet standards setting that promotes principles of openness, transparency, innovation, and economic competition. The final section of this chapter shifts attention back to IPv6, examining the implications of the slow transition and the reality of a diminishing store of Internet addresses.

## Political and Economic Implications of Protocols: A Framework

As points of control over global information architectures, Internet protocols and other information technology standards can serve as a form of public policy established primarily by private institutions rather than by legislatures. Not every technical protocol has significant policy implications and, as such, different types of contexts raise different political and economic concerns. This section presents six ways in which technical standards potentially serve as a form of public policy: (1) the content and material implications of standards can themselves constitute substantive political issues; (2) standards can have implications for other political processes; (3) the selection of standards can reflect institutional power struggles for control over the Internet; (4) standards can have pronounced implications for developing countries; (5) standards can determine how innovation policy, economic competition, and global trade can proceed; and (6) standards sometimes create scarce resources and influence how these resources are globally distributed.

## Standards and Substantive Political Issues

Technical protocols can have significant effects on substantive public interest issues that form the subject-matter of political debate. The history of IPv6 design and adoption policies reflects many of the ways in which standards have substantive public interest implications, including the protection of individual privacy, the ability of citizens to access knowledge, and government services such as disaster response, national security, and critical infrastructure protection. Perhaps the most substantive question about the Internet Protocol is whether the IP address space will scale to support the ongoing growth of the Internet.

The privacy question underlying the IPv6 address design is also illustrative of how Internet protocols can serve as a form of technological regulation similar to traditional law. In designing how a computing device such as a laptop would generate an IPv6 address, Internet engineers made decisions potentially impacting the privacy of a user's identity and geographical location while using the Internet. One option for generating IPv6 addresses, depending on technical circumstances, involved embedding a computing device's hardware serial number into the IP address. This technical decision would have created an environment in which online information could potentially be traced to a specific computer and therefore an individual's identity, and might also reveal information about an individual's geographical location. Internet engineers chose to architect some privacy protections into the design of IPv6 addresses. Although privacy advocates, particularly in the European Union, have raised more recent concerns about IP address privacy generally, this example of Internet engineers designing privacy features into protocols is indicative of how technical standards can influence civil liberties online. Similar privacy decisions arise in the development of encryption protocols and electronic health care information standards that determine the privacy and access afforded to individuals during the recordation, storage, and exchange of these medical records.

Privacy decisions enter protocol design in two ways. In cases such as IPv6, designers can engineer, or choose not to engineer, privacy protections into a protocol designed for some other purpose than privacy but that, in its design, raises privacy concerns. Designers or advocates can recognize that a new protocol creates potential privacy concerns and determine whether to reengineer the protocol to mitigate these concerns.

In other cases, such as encryption protocols, the objective of the protocol itself is to provide user privacy. As such, cryptography is an especially politically charged area of standardization designed to ensure the privacy of sensitive information such as financial, medical, personal, and national security data but also potentially serving as an impediment to military, diplomatic, intelligence, and crime fighting activities. Encryption protocols are often an area of conflict between individual civil liberties and government functions of law enforcement and national security. These protocols protect information privacy by mathematically manipulating data according to a predetermined algorithm called a cipher, also known as an encryption key. Transmitting and receiving devices use this encryption key to encode and decode data. Because criminals, hackers, or adversarial nations can use encryption protocols to mask information, governments have imposed restrictions on cryptography, including banning encryption outright, requiring licenses, or imposing export restrictions based on the strength of the encryption protocols. Prior to 1996 the US government grouped cryptography in the category of munitions, placing encryption under the requirements of the US International Traffic in Arms Regulations (ITARS). Companies could not export encryption products without a license and were prohibited from exporting to select countries including Cuba, Iran, Iraq, and Syria. The law essentially categorized encryption software companies as arms dealers. Later regulatory modifications reduced encryption restrictions, but the area of cryptography protocols illustrates how technical specifications mediate between conflicting social requirements such as the protection of individual privacy and the ability of governments to gather intelligence or perform necessary law enforcement and national security functions.

National security is a sphere in which contemporary societies ascribe particular—and usually exclusive—responsibility to the government. This is also a sphere in which technical standards play a significant and increasing role. Recall that, in the United States, the Department of Defense was the first institution formally interested in transitioning to IPv6. In 2003, the DoD's chief information officer announced that the Defense Department would transition to IPv6 by 2008. In the context of US efforts to prevent another terrorist attack like September 11, 2001, and to fight battles in Afghanistan and Iraq, the effective implementation of IPv6 was then viewed as a means for achieving "net-centric operations and warfare" that used distributed technologies and internetworked sensors to fight more efficiently and effectively.

Some technical protocols can improve national security or the ability to fight terrorism. Alternatively, the mere promise of new protocols can serve to reinforce political legitimacy, provide the impression of "doing something," reassure the public, or serve as a deterrent. In either case the protocol has a political role.

Advocates of IPv6, including governments seeking to advance national IPv6 strategies, have also linked the standard with political objectives. As chapter 4 described, the Japanese government, encouraged by Japanese corporations with a stake in IPv6 adoption, suggested that Japan's IPv6 mandate and corresponding industry product innovations would improve Japan's economic competitiveness in the wake of long-term stagnation. European Union policies linked IPv6 adoption with its Lisbon objectives of becoming the world's most competitive knowledge-based economy. The Korean government followed Japan in arguing that IPv6 expertise could make the country an "Internet powerhouse" and experience a corresponding reduction in unemployment and rise in GDP. Strategies about a future upgrade and promises about what it might accomplish not only reflected, but also strengthened political reputation or ideology by providing, through concealed and complex protocols the general public could not understand or even see, the impression that a solution existed. The history of IPv6 demonstrates how various groups can make use of the same technology as a resource for achieving many objectives.

Conversely, *problems* with standards can cause public safety concerns, economic harm, and loss of faith in political authority. There can be both economic and public safety consequences of using coexisting but distinct standards, as historian Ken Alder explained in his description of the loss of the NASA satellite *Mars Climate Orbiter*. One group of engineers had used the metric system while another team had used the traditional English system of measurement. This mathematical incompatibility, manifested in a software miscalculation, resulted in a 6-mile trajectory error and loss of the satellite. Web browser incompatibility reportedly prevented some Hurricane Katrina victims in the United States from registering for FEMA aid online—only victims using Microsoft's Internet Explorer could initially access FEMA's online registration. This incident followed reports of software incompatibility during the rescue and victim identification efforts immediately after the 2004 Southeast Asian tsunami. Various Thai agencies and organizations were unable to exchange documents because of incompatible proprietary document formats.[5] Incompatible technical standards that encumber such government services raise questions of particular political concern.

In addition to disaster response, protocol vulnerabilities can create opportunities for critical infrastructure attacks and general service outages in information networks that operate water control systems, electrical grids, financial markets, and air traffic control systems. The Internet itself has a generally distributed and redundant architecture seemingly making it difficult to disrupt. But a terrorist attack on the underlying power grid or the Internet's domain name system is always a concern. The DNS is centralized in that there must be a single root for the hierarchical name space. Despite the distribution of the DNS across numerous root servers, attacks have occurred, particularly distributed denial of service (DDoS) attacks. These attacks suspend the availability of websites by inundating them with requests, launched simultaneously from numerous unwitting computers. This approach is analogous to thousands of individuals simultaneously flooding an emergency dispatch system to prevent legitimate calls from reaching an emergency operator. The most notorious problem occurred on October 21, 2002, when a DDoS attack flooded the DNS root servers.[6]

Because the DNS is such a critical Internet function, protocols related to the DNS often have particular political and national security importance. Internet engineers proposed the DNS Security Extensions (DNSSEC) protocol to secure the critical Internet functions of root zone management and name and address resolution.[7] To illustrate the political complexity of Internet protocols, some view DNSSEC as not only a security protocol but also an opportunity to solve the question of who should have authority over the Internet's root zone file, one of the most divisive issues historically

5. See Berkman Center for Internet and Society, Open ePolicy Group's "Roadmap for Open ICT Ecosystems," September 2005. Accessed at http://cyber.law.harvard.edu/epolicy/roadmap.pdf.

6. For more information on distributed denial of service (DDoS) attacks and about the history of these attacks, see Laura DeNardis, "A History of Internet Security," pp. 694–695 in *The History of Information Security: A Comprehensive Handbook*, Karl de Leeuw and Jan Bergstra, eds., Amsterdam: Elsevier: 2007.

7. See generally, Brenden Kuerbis and Milton Mueller, "Securing the Root: A Proposal for Distributing Signing Authority," Internet Governance Project White Paper, May 2007.

confronting global Internet governance. Milton Mueller and Brenden Kuerbis have suggested that DNSSEC, though designed to make the Internet more secure, can provide a new solution for breaking the US government's legacy control of the root by using this new protocol to revise the root zone management procedures.

Critical infrastructure attacks that exploit protocol vulnerabilities or use DDoS techniques also have political consequences when they disrupt routine government services. In 2007, after Estonia removed a Soviet military monument from its capital, weeks-long denial of service attacks targeted and crippled the functionality of some of Estonia's state (and private) websites.[8]

Internet engineers long ago recognized the possibility of such security problems. In the late 1980s and early 1990s as Internet security breaches began to occur, engineers began to actively design security features into protocols. These concerns also led to Internet engineers closely linking the use of an existing security protocol, IPsec, with IPv6. The IPv6 specification originally mandated IPsec inclusion, a decision IPv6 advocates would later cite to promote IPv6 as more secure than IPv4. IPv6 advocates, ranging from the US Department of Defense, Japan's IT Strategy Council, and various IPv6 Forums, cited enhanced security as one rationale for upgrading to IPv6. Even testimonies in the 2005 US congressional hearings on IPv6 claimed that the protocol improved security.

The historical association of IPv6 with a separate security protocol (IPsec) is partly responsible for these claims of improved security, but as chapter 4 described, claims that IPv6 is more secure than IPv4 are contestable for several reasons. The exclusive linking of IPsec encryption with IPv6 in rationales for upgrading is somewhat misleading because IPsec can also accompany IPv4. It can also be argued that mixed IPv4 and IPv6 network environments are actually less secure than IPv4-only, especially when interoperable through translation gateways or protocol tunneling techniques. Various Computer Emergency Response Teams in the United States and around the world have identified numerous, intrinsic security vulnerabilities in IPv6 products. While not atypical of evolving protocols, the spate of IPv6 security weaknesses appears to weaken arguments that IPv6 is intrinsically more secure than IPv4. US government studies of IPv6 have also warned of security risks as a significant transition consideration for federal agencies.

8. See a description in the article "A Cyber-Riot," in *The Economist*, May 10, 2007.

This section has used IPv6 and several other protocols to illustrate some ways in which technical protocols make decisions directly related to issues of substantive political concern. This intersection between protocols and public policy falls roughly into three categories: (1) the ways in which protocols serve as a form of regulation making values decisions such as the extent of user privacy on the Internet; (2) the extent to which protocols enable, or impede, the ability of governments to perform paradigmatic functions such as national security and disaster response; and (3) the role protocols play in the functioning of critical information infrastructures such as financial systems, water control systems, and obviously the Internet.

### Standards and Political Processes

Technical standards can also more directly enter political processes when involved in formal systems of political authorization and representation, the electronic archival of government and public documents, and more informal conditions under which citizens engage in the public sphere. In democratic societies in particular, standards related to electronic voting machines and electronic voter registration supply a principal example. Engineering values such as security and transparency into the digital technologies increasingly underlying formal democratic processes is necessary for legitimacy and civic trust in government. For example, vote tabulation processes have historically been available for public scrutiny, with volunteers gathering in a room scrutinizing election ballots. The question of whether standards for electronic voting tabulations and information exchange are also open for viewing, as well as in a format that can be readily inspected, raises political concerns and provides a clear example of how emerging technical standards have implications for formal systems of authorization and representation.[9]

Beyond formal political processes, protocols affect conditions in which the public engages in broader political processes. Political engagement extends beyond voting rights to include equal opportunities for citizens to understand choices under consideration. For example, the electronic archiving of public documents is a fundamental responsibility

9. See, for example, Rebecca Bolin and Eddan Katz, "Electronic Voting Machines and the Standards-Setting Process," 8 *Journal of Internet Law* 3 (2004). Accessed at http://ssrn.com/abstract=945288. Also see Jason Kitcat, "Government and ICT Standards: An Electronic Voting Case Study," *Information, Communication, and Ethics in Society*, 2004. Accessed at http://www.j-dom.org/files/Kitcat-evoting_case.pdf.

of democratic governments and public access to these documents is essential for government accountability and for public deliberation over the efficacy of government institutions and policies. The ability of citizens to access electronic government documents significantly affects their capacity to participate in and critique public decisions. It is impossible to engage in successful public debate or reasoned critique of government action without firm knowledge of the content and implications of those actions.

The technical standards underlying electronic public documents, usually called document file formats, raise political concerns if they prevent governments from ensuring that electronic government archives are accessible in the future or if they are stored in a proprietary format that restricts user software choices in accessing documents, or if the standard locks public records into a format that is dependent on a single corporation to maintain. These concerns led the Commonwealth of Massachusetts, in 2004, to establish a policy to procure information technologies based on publicly available standards developed by an open technical community. The rationale for this "open standards" policy was economic, technical, and political: open standards would allow multiple vendors to compete and therefore reduce costs; they would enable technical interoperability; they would also provide user choice and prevent government information from being locked in a proprietary format dependent on a single vendor.

This open standards procurement policy, one of the first of many that governments would propose, was at the time a controversial strategy. The most contentious part of the Massachusetts announcement was its specification of Open Document Format (ODF) as the preferred document file format.[10] Also referred to as OpenDocument, ODF is an XML-based document file format for office applications such as word processing documents, spreadsheets, and presentations.[11] The new policy was controversial because the selection of this standard meant that the Massachusetts government would no longer rely on Microsoft Office applications, which

10. Information Technology Division of the Executive Office for Administration and Finance, Commonwealth of Massachusetts, Enterprise Open Standards Policy (Policy ITD-APP-01), January 2004. Accessed at http://www.mass.gov/Aitd/docs/policies_ standards/openstandards.pdf.

11. The Organization for the Advancement of Structured Information Standards (OASIS) approved the ODF specification in May 2005, and assumed responsibility for maintaining and updating the technical specification. The International Organization for Standardization (ISO) and the International Electrotechnical Commission (IEC) approved Open Document as an international standard (ISO/IEC 26300) in 2006.

were then based on proprietary standards for text, spreadsheet, and presentation documents. The formatting structures underlying office products like Microsoft Office have historically been proprietary—unpublished specifications not available for other vendors to create competing, interoperable software products. Rather than continue to use proprietary structures, the Commonwealth selected the ODF specification, which was available for anyone to access for free and implement on a royalty-free basis.

The Commonwealth's rationale for adopting ODF included its concern about the potential implications of giving a single corporate interest, Microsoft, the ability to control or limit access to public documents through proprietary formats and intellectual property restrictions. In a public statement about the importance of open standards in the context of the government's obligations to provide long-term accessibility to public records, Massachusetts employee Eric Kriss argued:

It should be reasonably obvious for a lay person who reflects on the concept of public records that the government must keep them independent and free forever. It is an overriding imperative of the American democratic system that we cannot have our public documents locked up in some kind of proprietary format, perhaps unreadable in the future, or subject to a proprietary system license that restricts access.[12]

The Massachusetts ODF decision, on the surface a recommendation involving an esoteric technical standard, attracted considerable attention, including a strong reaction from Microsoft, which had an obvious economic stake in retaining the large installed base of Office products in Massachusetts and which was in the process of introducing a more open document file format called Office Open XML (OOXML or Open XML), based on an XML document standard rather than the proprietary binary formats underlying previous versions of Office.[13]

Following a series of resignations, administration changes, and mounting political pressure, Massachusetts agreed that OOXML would also be an acceptable open format. This is just one brief story about standards related to public documents, and one that incorporates economic concerns and

12. Eric Kriss, secretary for the Executive Office of the Administration of Finance for the Commonwealth of Massachusetts, *Informal Comments on Open Formats.* January 14, 2005. Accessed at http://consortiuminfo.org/bulletins/sep05.php.

13. Microsoft's Open XML format was approved by the standards consortium Ecma International, which would make the standard freely downloadable from its website, and ultimately ratified as an international standard by the International Organization for Standardization.

corporate interests. But this is also an example of how politics and technical standards collide and the role technical standards play in the ability of citizens to hold governments accountable through open access to electronic government archives.

Particularly in liberal democracies, political participation extends further than formal systems of voting and government accountability to the public into more informal interactions of civil society and culture.[14] The advent of nonproprietary technical standards has increased society's ability to interact, produce information, and participate in the global public sphere, all characteristics that traditionally contribute to political and cultural discourse. Conversely, technical specifications that contribute to the ability of governments to filter, block, censor, or engage in surveillance of information flows can repress the public's ability to engage in political processes and participate in the global public sphere.

### Standards as a Site of Institutional Control over the Internet

Conflicts over the selection of Internet protocols, as well as the oversight of critical Internet resources have reflected institutional struggles for control of the Internet in the context of Internet globalization. In the early days of the Internet and its predecessor networks, Internet participants were both users and developers. These user-developers shared educational and cultural commonalities and primarily participated within American academic, research, and military contexts. They were trusted insiders with close familiarity with other insiders. Relative to later Internet contexts, access was limited. Enormous amounts of money were not at stake and there was no obvious linkage between corporate profits and standards development. No outsiders participated. In this collegial, relatively closed environment, standards consensus was uncomplicated and security was not a significant concern. The commercialization and international expansion of the Internet into businesses, across the globe and into homes heightened economic stakes, cultural complexity, and security concerns and transformed the prevailing trusted insider development environment into a more diffuse collaboration among strangers, including involvement of those not directly contributive to technical standards and those with pronounced corporate or political stakes in architectural outcomes.

The history of IPv6 demonstrates how the breakdown in trusted insider status and the globalization of the Internet transformed the Internet

14. See, for example, Jack Balkin, "The Constitution of Status," 106 *Yale Law Journal* 2313 (1997).

architecturally and administratively. First, the 1992 "Kobe affair" reflected anxiety about outsiders influencing architectural decisions and led the Internet standards community to solidify and articulate more open Internet standards governance approaches. In the context of increasing Internet internationalization and commercialization, the Internet Architecture Board (IAB) responded to concerns about Internet address space exhaustion by taking an uncustomary step of proposing a specific OSI-based protocol to replace IPv4. The IAB had recently become associated with a new umbrella organization, the Internet Society, that in several characteristics broke with historical traditions in Internet standards development. The Internet Society cultivated links with international standards bodies, received direct corporate funding, promoted formal membership, and was equipped to respond to the emerging threat of lawsuits related to protocol development.

IETF participants were dismayed by the IAB's recommendation for several reasons. The IAB seemed to be relinquishing responsibility for Internet standards development and change control to the international standards process of ISO. The decision disseminated from a top-down mandate without the benefit of open hearings and public review in contrast to the IETF's historical bottom-up decision making process. Some Internet insiders also believed the recommended standard to replace IPv4 was untested and expressed concern about undue corporate influence. They believed the IAB lacked the legitimacy it once garnered because participants were no longer ARPANET veterans or those directly involved in development and coding. IETF participants no longer viewed the IAB as trustworthy insiders concerned with preserving standards-setting continuity and traditions. This breakdown in trust resulted in a solidification and articulation of the standards community's operating philosophy of grassroots, consensus-based, and open standards development and a rejection of top-down mandates.

The very choice of the next generation Internet protocol was an issue of selecting what institution would have authority to establish the Internet's architectural directions. While the assessment process emphasized that only technical considerations would influence protocol selection, a salient consideration appeared to involve which standards-setting community would retain or gain architectural control. The ISO-related alternative had considerable momentum: endorsement by most western European governments, investment by some prominent American technology companies, and US government support because of its endorsement of the GOSIP architecture. If the IETF had selected the ISO-related protocol, it would have raised complicated questions about which standards institution would have change control over the protocol in the future. In other words, ISO would suddenly have greater control over the Internet's architecture. The selection of SIPP, an extension of the prevailing IPv4 standard, served to secure the power of the Internet's existing standards-setting establishment to control the new protocol. A decade after this decision, tension between other standards organizations and the Internet's traditional standards body, the IETF, still existed. Recall the ITU's proposal to take a greater role in Internet standards setting and the suggestion of the UN Working Group on Internet Governance (WGIG) that standards should be the purview of governments. These conflicts over standards control reinforce how standards selection is also a question of institutional power selection.

### Pronounced Implications of Standards for Developing Countries

A recurrent theme throughout the history of IPv6 has been the promise of this new protocol to promote economic growth and social progress in the developing world. Internet engineers, in 1990, identified the need for a new standard to expand the address space to meet the anticipated requirements for Internet connectivity in emerging markets and elsewhere. They also proposed the distribution of IP addresses to international registries that would support developing countries. The recommendations of the Working Group on Internet Governance similarly cited the needs of the developing world to bolster its case for diminished American power over IP address administration and other Internet governance functions. The belief that Internet governance legitimacy required participation of developing countries was the premise of this argument.

The UN-sponsored World Summit on the Information Society more generally identified standardization as a foundation for the global information society and cited open standards as a precursor to affordable information and communication technology diffusion in the developing world:

Standardization is one of the essential building blocks of the Information Society. There should be particular emphasis on the development and adoption of international standards. The development and use of open, interoperable, nondiscriminatory and demand-driven standards that take into account needs of users and consumers is a basic element for the development and greater diffusion of ICTs and more affordable access to them, particularly in developing countries.[15]

15. WSIS Declaration of Principles, "Building the Information Society: a Global Challenge in the New Millennium," Geneva, December 2003. Accessed at http://www.itu.int/wsis/docs/geneva/official/dop.html.

In all these cases, standards were viewed as a tremendous opportunity for developing countries. However, the institutional processes and underlying intellectual property arrangements of technical standards do not always reflect developing country interests. Intellectual property restrictions can have pronounced effects on developing countries. As chapter 4 described, the historical traditions of the IETF have encouraged minimal intellectual property restrictions on Internet protocols. This approach has enabled individuals and companies to use these open protocols to develop competing, interoperable products. Many scholars cite this minimization of intellectual property restrictions on core Internet protocols as a major contributor to the rapid innovation and growth of the Internet. However, TCP/IP and other core networking standards are only a fraction of the standards required for communication over the Internet. The number of standards required for information exchange has significantly increased to support multimedia applications and versatile devices that embed numerous standards for voice, video, text messaging, and other applications. Many of these standards have intellectual property restrictions. Another complexity is that there is an enormous diversity of intellectual property policies among standards-setting institutions, even those in the same industry.[16] Furthermore, intellectual property rights are not necessarily disclosed until long after a standard is widely deployed.

These complex circumstances can disadvantage entrepreneurs in developing countries who have not necessarily been involved in the development of standards and who may not have large patent portfolios. They are usually later market entrants who were not involved in the development of a standard they plan to implement in their products. Implementation of a standard can require permissions, so emerging companies wanting to implement a standard are dependent on these permissions. It also may require a royalty payment and legal expertise to deal with licensing complexities. The lack of disclosure of intellectual property rights furthermore raises the specter of investing in product development (based on technical standards) only to face a patent infringement lawsuit at a later time. The implication is that the patents and lack of patent disclosure underlying technical standards create impediments to later entrant entrepreneurs in developing countries seeking to compete or innovate in technical standards relative to large multinational companies.

16. Mark A. Lemley, "Intellectual Property Rights and Standard-Setting Organizations," 90 *California Law Review* 1889 (April 2002). Accessed at SSRN: http://ssrn.com/abstract=310122.

Standards attorney Andrew Updegrove summarizes the intellectual property advantages of large multinational corporations. They have extensive research and development capacity, they have large legal staffs to deal with intellectual property rights, they regularly engage in cross-licensing agreements with other large companies, they are culturally well-versed in the historical traditions of standards-setting institutions, and they have enormous patent portfolios.[17] Updegrove describes the accompanying phenomenon of "standards-based neocolonialism":

> Royalty bearing patent claims are embedded in the standards for products such as DVD players and cellular phones. If the royalties are high enough, the patent owners can have such products built in emerging countries using cheap local labor, and sell them there and globally under their own brands. Meanwhile, emerging company manufacturers can't afford to build similar products at all.[18]

Having to pay royalty payments, as well as facing the difficulties of addressing standards-based intellectual property complexities, can discourage new entrepreneurial activity. Information technology development and entrepreneurship hold promise to provide economic opportunities in new markets, but emerging companies in the developing world have distinct disadvantages that heighten the effects of intellectual property restrictions.

Developing countries also face barriers to participation in standards setting. The IETF has traditionally been considered one of the most open standards-setting institutions because any interested party has been able to participate. The legitimacy of Internet standards development has always derived, in part, from this open institutional approach. Other standards-setting institutions are not necessarily this open and have membership requirements, impose fees, or may be closed to new members entirely. In some institutions, developing country interests have little voice in the process dominated by private multinational corporations. While large corporations have experience with the historical traditions of standards-setting institutions, corporations in developing countries may not. As the number of standards required to meaningfully use the Internet has increased, so has the number of organizations that establish these

17. Andrew Updegrove, "It's Time for IPR Equal Opportunity in International Standards Setting." 6 *Consortium Standards Bulletin* (August–September 2007). Accessed at http://consortiuminfo.org/bulletins/aug07.php#editorial.
18. Andrew Updegrove, "Government Policy and Standards-Based Neocolonialism." 6 *Consortium Standards Bulletin* (August–September 2007). Accessed at http://consortiuminfo.org/bulletins/aug07.php#feature.

standards—and so has the number of different policies regarding institutional participation. This book has described how technical protocols can establish public policy. If developing country interests do not enter the standards-setting process, their interests are not directly reflected in this policy-making process.

Developing countries' information technology industries are impacted by standards-related intellectual property restrictions as described, but developing countries are not only technology developers; they are also technology users. The costs and restrictions of standards-related intellectual property rights are transferred to users. This creates a heightened burden on developing countries as they often lack the installed base of information technology infrastructures inherent in more developed regions. They often must purchase essential products from Western companies, even if the products are manufactured in their countries. For example, the international community has placed great emphasis on the promise of broadband wireless technologies like WiMAX, Wi-Fi, and GSM to cost effectively deliver critical infrastructures in the developing world. WiMAX, short for Worldwide Interoperability for Microwave Access, is an emerging broadband metropolitan wireless technology that has a further transmission range than other wireless Internet access mechanisms like Wi-Fi. Wi-Fi actually refers to a set of wireless Internet standards developed by the IEEE. GSM, or Global System for Mobile communications, is a popular cellular networking standard. In the United States, for example, these are often viewed as access technologies, but in developing countries without existing fiber and copper backbone infrastructures, broadband wireless can serve as both the backbone infrastructure and as access mechanisms. Wi-Fi, WiMAX, and GSM are standards. Furthermore, they are standards with embedded intellectual property rights, in the form of patents, that raise the prices of these broadband technologies. Emerging markets without an installed base of existing products disproportionately bear the marginal cost increases engendered by this escalation of embedded intellectual property rights.

IPv6 and other information technology protocols promise to improve access and economic opportunities in the developing world. However, the degree of openness in standards has pronounced implications on developing countries in several ways: intellectual property restrictions impede the ability of later entrant entrepreneurs in developing countries to create cost-effective products; standards-based intellectual property restrictions also heighten product costs, which have greater effects on areas that lack a significant installed base of information technology infrastructure; and developing country interests often do not enter the standards-setting process, because of either institutional barriers or later entry.

## Standards, Innovation, and Competition Policy

Standards inherently have relevance to innovation and competition policy. The openness of a standard either promotes or impedes free trade and either promotes or obstructs competition. The objective of Internet protocols is to create global interoperability and therefore encourage free trade, global information flows, and competition among information technology companies. Prior to the development of TCP/IP, the proprietary network protocols underlying one technology company's product line did not necessarily communicate with the proprietary protocols used within another technology company's product lines. The availability of openly published protocols such as TCP/IP enabled multiple companies to create competing interoperable products.

However, the standards-selection process itself often involves tremendous conflicts among competing businesses. In the selection of the next generation Internet protocol, the decision reduced to a choice between an OSI-based protocol, TUBA, versus a protocol more closely related to the existing version of the Internet Protocol. DEC had invested heavily in OSI and was backing the selection of TUBA. Recall that a prominent Internet engineer who worked for DEC on issues related to OSI, and specifically to introducing OSI in the Internet, was a member of the IAB that made the initial, controversial decision to recommend an OSI-based protocol as the next generation Internet protocol. On the other hand, Sun Microsystems was heavily invested in TCP/IP protocols and had an interest in more closely aligning the new protocol with existing TCP/IP standards. The selection internalized many variables, but the competition between DEC and Sun Microsystems was certainly one factor.

A more complicated economic area related to standardization is global trade. When such exchanges—and the technical standards that govern them—directly affect diplomatic relations and national economic competitiveness, they implicate a core governmental function. Many countries, and in particular China, have established well-defined national standards strategies that invest in standards development and specify adoption policies based on various criteria. This has certainly been the case with China's national IPv6 strategy. The United States has been one of the few large countries with a more market-based standards strategy. Among the countries with strong national standards policies, a major rationale for direct government policy-making and intervention is the objective to become more competitive in global information technology markets.

The World Trade Organization (WTO) Agreement on Technical Barriers to Trade (TBT) recognizes the important role standards play in the facilitation of international trade and asserts that standards should not create unnecessary obstacles to trade. However, intellectual property rights in standards can inhibit the adoption of international standards and the development of products based on these standards. Christopher Gibson, in "Technology Standards—New Technical Barriers to Trade?" argues that standards are increasingly emerging as non–tariff barriers (NTBs) and cites WAPI (Wireless Local Area Network Authentication and Privacy Infrastructure), the Chinese national standard for wireless LAN encryption, as a case study in this area.[19] China announced that its proprietary encryption protocol for Wi-Fi networks would be a nationally mandated standard. Foreign equipment manufacturers wishing to sell in China's lucrative wireless local area network (LAN) market would have to license the proprietary protocol, WAPI, from one of China's indigenous network equipment manufacturers, effectively creating a barrier to trade. China also has raised similar issues about standards as global trade barriers to the WTO committee on technical barriers to trade. Although the trend in the emerging global knowledge economy has been to lower traditional barriers to global trade, proprietary standards are increasingly emerging as alternative technical barriers to trade.

### Standards Create Finite Resources

Many technical standards create and allocate the finite resources required for access to information networks. How these resources are distributed, and by whom, can raise political questions and issues of distributive justice, as well as economic concerns. Some standards partition and allocate radio frequency spectrum among users (e.g., broadcast standards, Wi-Fi, and cellular standards). Others prioritize the flow of information over a network based on application, such as prioritizing voice applications and decelerating peer-to-peer video. Other standards divide up orbital slots in satellite systems. Some standards assign rights of access to local broadband services. Still others, such as DSL, provide an asymmetric distribution of bandwidth whereby downstream communications to users are privileged over upstream communications from users to the network. The question of who controls

19. Christopher Gibson, "Technology Standards—New Technical Barriers to Trade?" 2007. Accessed at http://ssrn.com/abstract=960059.

the scarce resources created by standards is a critical one. In a global Internet economy, control over standards and the scarce resources they create increasingly determines wealth. Citizens in countries with greater control over standards development and adoption and the scarce resources created by those standards have distinct economic advantages and opportunities to produce and control information.

The Internet address space under IPv4 created 4.3 billion unique addresses. The initial distribution of these technical resources involved allocation to those organizations involved in the early development and adoption of Internet predecessor networks. American institutions received large blocks of addresses when the Internet was primarily used in the United States. These initial allocation circumstances, along with rapid Internet user growth and new applications like voice over the Internet and widespread wireless access, raised concerns that the Internet address space would be depleted and that the rapid growth of the Internet was unsustainable. The selection of IPv6 was obviously designed to address this concern and sought to expand the number of available addresses to 340 undecillion addresses.

The question of who should control and allocate IP addresses has been a mounting controversy in Internet governance. Centralized control has historically existed in part to maintain the architectural principle of globally unique addresses. Address assignment stewardship shifted from a single individual to an American institutional framework to a regionally distributed structure (RIRs), though still with centralized coordination. These regional Internet registries are primarily nongovernmental bodies. They are nonprofit corporations with paying memberships. RIR member institutions consist primarily of ISPs, telecommunications companies, and other large corporations in each respective region. One ongoing public policy question examines the implications of these organizations, ultimately controlled by private companies, controlling the vital resources required for Internet participation.

This book has described how oversight of the entire address reserve, including the allocation of address resources to international registries, has remained somewhat centralized under ICANN. The question of who should centrally administer the allocation of resources has remained a source of controversy centered around issues of international fairness, institutional legitimacy, security, and stability. This question is at the heart of ongoing conflicts between US oversight of the IANA function under ICANN and the possibility of turning that function over to the United Nations or other organization perceived as more international and multistakeholder. If history is any indication, this impasse will remain for the foreseeable future.

## Values in Protocols

Protocols are often invisible to end users. The general public may not be aware of protocols, never mind aware of their political implications or how or by whom they are established. Internet users who are familiar with technical protocols may be unconcerned with the standards-setting processes that create them. Many view Internet standards as working because of the technical interoperability they achieve and view them as objective technical design decisions.

Others view Internet standards governance as an inherently democratic process that is decentralized, open, and participatory. Many scholars view Internet standards setting as a paragon of democratized technological design, at least historically. Legal scholar Larry Lessig describes how control over the Internet's architecture has shifted from a collaborative technical community designing values of personal and technical freedom into the Internet to hegemonic corporations formerly threatened by the Internet but now the "invisible hand, through commerce . . . constructing an architecture that perfects control."[20] Dominant corporations, according to Lessig, have replaced the collaborative and open efforts of the Internet user community, and dictate architectural directions in a manner that threatens innovation and the foundational freedoms and values of the Internet.

The history of IPv6 suggests that early standards development also reflected conflicts among competing companies and institutional struggles, and encountered obstacles to completely open procedures. IETF working groups are open to public participation and an excellent example of collective action in the technical sphere, but even such an open approach has participatory barriers. Impediments to democratic participation include the four horsemen of money, access, culture, and knowledge. Because involvement in developing standards like IPv6 is uncompensated activity, participants usually have the financial backing of salaries from corporate employers supporting their participation. Within the IETF, individuals have "one voice" from which to participate but the individuals also represent the interests of the institutions funding their involvement. Most communications occur over the Internet, requiring access, and

20. Lawrence Lessig, *Code and Other Laws of Cyberspace*, New York: Basic Books, 1999, p. 6.

there are cultural norms such as communicating exclusively in English and subscribing to the procedures and values the IETF espouses. Participation in network standards work also requires technical understanding of abstract and complex protocol issues, an obvious barrier to general public participation.

Standards communities often view their decisions as based primarily on technical criteria, while outsiders view Internet standards governance as a democratic process. An inherent contradiction underlies the standards-setting formulation as characterized by the IETF and described by outsiders. The IETF specified that only technical considerations would factor into the IPv6 selection process. The IETF process also embraces a one voice, one vote process, and rough consensus. The belief in technical neutrality denies the role of a political process in standards setting. The inherent contradiction is that the process cannot be apolitical, on one hand, and a one voice, one vote and consensus-oriented political structure, on the other. Addressing questions about standards setting requires deciding whether the process is political. This book has examined how standards decisions are not unadulterated technical formulations but reflective of political and economic exigencies, warranting critical examination of the standards-setting process.

If technical protocols have political and economic implications, the issues of *who* decides in matters of standards setting and *how* they decide are salient questions. Technical protocols usually arise from the coordinated action of private actors or, if proprietary, from the market dominance of a private actor. Private industry engages in an essentially political process of establishing protocols with public interest implications. These institutions not only have market power to determine how innovation policy and global trade should proceed, but they acquire the power to make decisions influencing the rights of citizens who use the technologies based on these protocols.

The many organizations that establish Internet standards adhere to different procedural norms; exhibit disparate levels of participatory and informational openness and transparency; and have different policies on membership, due process, intellectual property rights, public document availability, and other key institutional criteria. This book focuses on IPv6 as a case study and therefore primarily on the institutional practices of the IETF, but this organization is only one of many organizations (e.g., IEEE, W3C, ITU, ISO, OASIS, ECMA) that establish Internet-related technical standards.

Technical expertise is the primary basis of legitimacy for standards setting, but because of the public policy implications of technical standards, additional legitimacy criteria must derive from principles of openness, transparency, and accountability. In practice, however, there are no consistent norms for how standards are set across different institutions, who may contribute to the standards development process, whether deliberations, proceedings, and documents are made publicly available, whether intellectual property is made available on a royalty free or other basis, and many other criteria. The procedures of some standards organizations are more susceptible to manipulation, giving preference to narrow interests rather than reflecting broader stakeholder interests. Some institutions exclude nonmembers, set standards in a nontransparent manner, and charge fees for specifications, all qualities that eliminate the possibility of direct multistakeholder participation or oversight. In this regard the IETF is one of the most open of all standards-setting organizations.

Normative standards of openness should be a common denominator within standards organizations and underpin any attempts to address best practices in technical standards setting. The various rationales for best practices based on openness include the objectives of promoting technical interoperability, encouraging competition and innovation, and attempting to reflect or be accountable to the public interest. The technical rationale is interoperability, which promotes the global flow of information by encouraging universally accessible standards rather than proprietary, balkanized standards. The economic rationale for best practices in standards setting based on principles of openness is both to promote product competition and innovation. Open standards, in theory, prevent anticompetitive practices and trade barriers and provide a more level playing field for entrepreneurs in developing countries to compete in global technology markets. Open standards furthermore promote Internet innovation by sustaining the Internet's tradition of encouraging royalty-free standards. The political rationale for best practices based on principles of openness is to enable effective and accountable government services, to reflect values of democratic access to knowledge, to provide greater openings for developing country interests to enter standards-based public policy decisions, and to create the necessary legitimacy for standards-setting processes to establish public policy.

### Conceptions of Openness[21]

Discussions about openness in standards setting are controversial and imprecise with a wide range of opinions on the meaning of both "openness" and "standards." Some definitions refer to open standards exclusively

21. This discussion of conceptions of openness and of a framework for open standards was first introduced in the Yale Information Society Project White Paper by

in terms of achieving technical interoperability between software or hardware products based on the standard, regardless of what company manufactured the product. Economic definitions of open standards are based on *effects*: effects on market competition, efficiency, improvements to national economic competitiveness, or the product innovation. For example, in "An Economic Basis for Open Standards," economist Rishab Ghosh suggests that open standards can be defined so as to promote full competition, and therefore innovation, among vendors developing products based on these open specifications.[22] Because of this desirable economic effect, Ghosh suggests that public procurement policies should promote open standards.

Neither purely technical nor economic definitions take into account the reality that technical standards embody values and, once developed, have political consequences. Other conceptions of openness are more expansive and address political questions about how standards-development processes can reflect procedural values of openness and transparency or how standards, once developed, maximize user choice, promote democratic freedoms, or enable effective government services. Open standards definitions originate in all sectors—industry, standards-setting organizations, governments, and academe. This section will describe some of these various approaches; the following section will synthesize these approaches into a framework for best practices in Internet standards governance that aims to promote interoperability, encourage competition and innovation, and take account of the political implications of standards.

Many governments have recognized the technical, economic, and political concerns raised by protocols and have taken an interest in defining and promoting open standards. There have been numerous instances of governments establishing policies to procure products that adhere to principles of openness and interoperability. Countries as diverse as Australia, Brazil, India, China, Sri Lanka, Belgium, Croatia, Hong Kong, and Malaysia have introduced open standards policies, often called government interoperability frameworks (GIFs).[23] The overarching purpose of these government policies

---

Laura DeNardis and Eric Tam, "Open Documents and Democracy: A Political Basis for Open Document Formats," November 2007. Available at http://papers.ssrn.com/sol3/papers.cfm?abstract_id =1028073.

22. Rishab Ghosh, "An Economic Basis for Open Standards," December 2005. Accessed at http://flosspols.org/deliverables/FLOSSPOLS-D04-openstandards-v6.pdf.

23. See UNDP note "Government Interoperability Frameworks in an Open Standards Environment: A Comparative Review." APDIP e-Note 20 (2007). Accessed at http://www.apdip.net/apdipenote/20.pdf.

is interoperability, the ability of government agencies to exchange information with each other and with citizens, with open standards cited as the primary method for achieving this interoperability.

For example, the Brazilian federal government issued an interoperability policy establishing the adoption of open standards for technology used within the executive branch of the federal government. The Brazilian model is representative of many open standards policies. It is limited to internal government communications and information exchanges with citizens and specifically states that the policies cannot be imposed on the private sector, citizens, or government agencies outside of the federal government, although it does request voluntary adherence to the standards specifications. The federal standards policies apply to new purchases and upgrades to existing systems rather than mandating a complete cut over to new products. Brazil, like other countries with open standards policies, cites a combination of technical, political, and economic justifications. Most policies express public service rationales such as improving services to citizens and avoidance of locking users into a single vendor's products, technical goals of seamless information exchange among agencies, and economic goals of lowering costs, promoting economic competition and innovation, and competing in global markets and exchanging information with global trading partners.

Brazil's definition of interoperability primarily addresses a standard's effects. Does it enable multiple, competing technologies? Does it create the ability to exchange information among heterogeneous technology environments? Does it provide users with product choice or result in single vendor lock-in? Principles of openness, choice, and technical heterogeneity underlie this definition.[24]

Other government interoperability frameworks have similar objectives but differ in their definitions of what constitutes an open standard. The European Union established a "European Interoperability Framework for Pan-European eGovernment Services" to promote Europe-wide electronic interoperability among public administrators, citizens, and corporations. The European framework defines an open standard as meeting the following minimum requirements: it must be developed in an open

24. Brazilian Government Executive Committee on Electronic Government, e-PING Standards of Interoperability for Electronic Government, Reference Document Version 2.0.1, December 2006 (translated by the Brazilian government). Accessed at http://www.governoeletronico.gov.br/governoeletronico/publicacao/down_anexo .wsp?tmp.arquivo=E15_677e-PING_v2.0.1_05_12_06_english.pdf.

decision-making process, the standard must be published and available either freely or at minimal cost, *and* the intellectual property (e.g., patents) "of (parts of) the standard is made irrevocably available on a royalty-free basis."[25] Many other open standards policies do not require that standards-based intellectual property be made irrevocably available on a royalty-free basis but may give preference to royalty-free standards when possible.

It is also notable that this definition includes openness criteria for a standard's development process rather than exclusively focusing on the standard's economic effects once developed. The development process must be open to anyone, maintained by a nonprofit institution, and embody democratically oriented criteria of transparency and a majoritarian or consensual decision-rule. The implication is that the standards development *process*, which might include public policy decisions, is as pertinent to definitions of openness as the material *effects* of a standard. Another distinguishing characteristic of this definition is the requirement that any underlying intellectual property be made irrevocably available on a royalty-free basis.[26] The Danish National IT and Telecom Agency also published a definition of an ideal open standard emphasizing that the standard must be documented in all its details and accessible to anyone free of charge with no discrimination and no payment as a condition to use the standard.[27]

Open standards advocates have also linked standardization issues directly to the Universal Declaration of Human Rights,[28] which the General Assembly of the United Nations adopted in 1948. A group of open standards advocates published The Hague Declaration on open standards, signed

25. IDABC Working Document, "European Interoperability Framework for Pan-European eGovernment Services," Version 4.2 9, January 2004. Accessed at http://ec.europa.eu/idabc/servlets/Doc?id=1674.

26. Many irrevocable royalty-free policies include protections such as reciprocity and defensive termination clauses. See, for example, Lawrence Rosen, "Defining Open Standards." Accessed at http://www.rosenlaw.com/DefiningOpenStandards.pdf.

27. See the Danish definition of an open standard accessed at http://www.itst.dk/arkitektur-og-standarder/files/040622_Definition_of_open_standards.pdf.

28. The Universal Declaration of Human Rights was adopted and proclaimed by United Nations General Assembly resolution 217 A (III) on December 10, 1948. The full text of the Declaration is available at http://www.un.org/Overview/rights.html.

eventually by thousands of individuals from government, industry, and civil society.[29] The Declaration cited several international rights and freedoms from the Universal Declaration of Human Rights, including the right to freedom from discrimination by government (Articles 2 and 7), the right to participate in government (Article 21.1), and the right of equal access to public services (Article 21.2). The Declaration summarized how government services, educational opportunities, and freedom of speech increasingly occur and are exercised in electronic spheres rather than in physical spaces. Open access to the Internet, enabled by free and open digital standards, is necessary to preserve human rights in these electronic spheres. Furthermore, the Declaration emphasized the role governments could play through exerting market influence in procurement policies and, by example, in encouraging open digital standards. The Declaration made the following three demands of governments: "procure only information technology that implements free and open standards; deliver e-government services based exclusively on free and open standards; and use only free and open digital standards in their own activities."[30]

The most contentious area of open standards definitions relates to standards-based intellectual property rights. The policies of some standards-setting organizations have asserted that intellectual property rights should be available under royalty-free terms but many also have adopted policies that the standard be available on a so-called "reasonable and nondiscriminatory" (RAND) basis. Mark Lemley's study, "Intellectual Property Rights and Standards-Setting Organizations," describes the diversity of approaches to how standards bodies treat intellectual property, but finds that RAND licensing approaches are the most prevalent.[31] Although RAND licensing approaches are well intentioned, their implementation can be problematic due to a lack of clarity over the meaning of both "reasonable" and "nondiscriminatory." Lemley notes that most organizations with RAND licensing requirements do not specifically define RAND.[32] Undefined variables include whether intellectual property rights holders are obligated to license to any entity or just to other standards body

29. The Hague Declaration was published by individual members of the Digital Standards Organization on May 21, 2008. Accessed at http://www.digistan.org/hague-declaration:en.
30. Ibid.
31. Mark Lemley, "Intellectual Property Rights and Standard-Setting Organizations," Boalt Working Papers in Public Law, Paper 24, 2002.
32. Ibid., p. 109.

members, what constitutes a reasonable royalty fee, and what constitutes reasonable and nondiscriminatory substantive licensing terms. In practice, the requirement for RAND licensing lacks a consistent or clear meaning—sometimes even within the same standards-setting organization.

Critics of RAND licensing practices also question whether the Internet would have experienced such growth in numbers, geographic scope, and technological innovation if its underlying protocols (e.g., FTP, HTML, HTTP, and IP) had been controlled by a single vendor or group of vendors under RAND terms rather than made available on a primarily royalty-free basis. As described earlier in this book, the IETF's policy is to, whenever possible, select protocols with no intellectual property restrictions. The World Wide Web Consortium, citing the objective of promoting ubiquitous adoption of web standards, has established a policy of issuing recommendations only if they can be implemented on a royalty-free basis, although there is a mechanism for allowing exceptions.[33] Ghosh has noted that royalty-free policies—which may conflict with defensive suspension clauses in F/LOSS (Free/Libre Open Source Software) licenses—may be too strict in some markets like mobile telephony and not strict enough for office applications. In the case of irrevocable royalty-free terms, such rules could produce undesirable results such as potentially excluding Adobe's PDF as an open standard because of its revocable royalty-free terms.

Other institutional definitions of open standards focus on the standards-setting process and issues of public participation, transparency, and accountability. The ITU has defined open standards as those that are "made available to the general public and are developed (or approved) and maintained via a collaborative and consensus driven process."[34] The ITU's openness definition also states that the standards-setting process should not be dominated by any one interest and should articulate a specification in detail sufficient to enable the development of heterogeneous competing products that implement the standard.

Some information technology companies also offer definitions of open standards. These companies have a vested interest in the definition, which they can use to evaluate and promote the degree of openness in their own products and use to criticize the openness compliance and possible anti-competitive practices of competitors. Sun Microsystems laid out open standards criteria in an open letter from its chief technology officer Greg Papadopoulos.[35] The letter recounts earlier, competing visions for the "information superhighway"—referring to set top box approaches backed by cable companies, studios, and Wall Street versus the early Internet, which thrived without enormous financial backing. Papadopoulos suggests that the Internet approach prevailed because, unlike other approaches, it had standards. Sun Microsystems recommends *Common Criteria for IT Interoperability* presenting minimal requirements a specification must meet to be characterized as an open standard. According to Sun, the development process must be democratic and collaborative, have well-documented processes, include commitments to disclose and license intellectual property rights, and provide for the actual standard specification to be open to public review at least once during its development lifecycle. The Sun recommendation on technical aspects of open standards emphasized high interoperability in information exchange among computer programs, including the ability "to use, convert, or exchange file formats, protocols, schemas, interface information or conventions, so as to permit the computer program to work with other computers and users in all the ways in which they are intended to function."[36]

Ken Krechmer's frequently cited paper, "Open Standards Requirements," expands the definition of open standards further to include not only economic effects resulting from an open standard's implementation and openness in the process of standards setting, but also the concept of openness in use.[37] Krechmer's requirements include openness criteria for development such as participatory openness, due process, and consensus. He also discusses requirements for the implementation of openness, including public availability of the standard and intellectual property arrangements that are not cost prohibitive, do not favor one competitor over others, and do not inhibit further innovation. Krechmer's definition addresses openness requirements directed at technology users, including choice of vendor implementation, ongoing support for the standard over the life of the product implementing the standard, and backward compatibility with previously purchased implementations.

35. See letter from Greg Papadopoulos, senior vice president and chief technology officer, on value of open standards, "Open Standards at Sun." Accessed at http://www.sun.com/software/standards/overview.xml.

36. Ibid.

37. Ken Krechmer, "Open Standards Requirements," 2005. Accessed at http://www.csrstds.com/openstds.pdf.

Open source advocate Bruce Perens further defines open standards by the *principles* he believes should underlie the development and adoption of technical specifications.[38] One of the principles Parens cites is maximization of user choice in that an open standard does not lock users into a single vendor's products. Institutions establishing open standards should not favor a particular vendor over other vendors. Perens also suggests that open standards should be ubiquitously available and capable of implementation on a royalty-free basis.

## Best Practices in Internet Standards Governance

For standards to reflect the public interest, open standards advocates and policy makers often recommend either greater government intervention in the standards process, the involvement of advocates in the standards process, or the promotion of greater public involvement. This book has challenged the efficacy of these possibilities for a variety of pragmatic and normative reasons.

Instead, the voluntary adherence to open procedures, processes, and effects (or adherence encouraged by public and private procurement practices) within standards-setting institutions can bolster the legitimacy of international standards setting. Various stakeholders are likely to continue to arrive at different, contextually specific definitions of open standards. This section will draw upon lessons and norms within the IETF and previous efforts to define and recommend open standards requirements to consider a definition of maximal openness. As the previous section explained, some definitions focus exclusively on the objective of technical interoperability; other definitions focus on the economic effects of openness; other definitions link open standards to human rights and democratic values of equal participation, accountability, and transparency. The following will stipulate a definition of maximal openness that takes all three spheres into account. In reality, it would be implausible to impose maximum openness equally in all contexts and circumstances, so the purpose is not to advocate the universal implementation of openness in Internet standardization but to fix the definition as one side of a spectrum of standards policy options, and as the criteria which could meet the following normative objectives:

• Technical rationale   Open standards promote maximum technical interoperability enabling the universal exchange of information among

38. Bruce Perens, "Open Standards: Principles and Practice," Accessed at http://perens.com/OpenStandards/Definition.html.

technologies made by different manufacturers. Universal and open interoperability standards, rather than proprietary standards or balkanized standards that differ by nation, product line, or geographical area, promote global access to knowledge.

• Economic rationale   Open standards enable competition among products based on the standards and provide a level playing field for product innovation. Standards prevent anticompetitive and monopolistic practices, promote rather than impede global trade, and provide openings for developing country entrepreneurs to compete in global information technology markets.

• Political rationale   Open standards-setting practices adhere to procedural norms and processes that maximize the legitimacy of standards institutions to make decisions that establish public policy in areas such as individual civil liberties, democratic participation in electronic cultural and political spheres, user product choice, and the ability of the public to access and disseminate knowledge. Standards enable efficient and accountable government functions such as systems of political authorization and representation, disaster response, national security, critical infrastructure protection, eGovernment services, and the archiving of public documents.

The following section will recommend a framework of best practices in Internet standards setting based on adherence to the normative principles listed above including participatory and economic openness, technical interoperability, universality, accountability, and transparency. Accordingly, the framework will define openness in three contexts: a standard's development, its implementation, and its use. The framework will include requirements of maximal participatory openness and transparency in development; the absence of hindrances to full competition and multiple competing implementations; and requirements of maximum technical interoperability among heterogeneous systems and therefore user choice.

Before examining the characteristics of a completely open standard, it is helpful to understand the criteria that anchor the opposite pole in standards policy choices—a completely closed specification. First, it is a misnomer to call a proprietary (or closed) specification a "standard" because this nomenclature implies some degree of coordination and use by multiple parties. A single company develops, owns, controls, and uses a proprietary specification. By definition, the company does not make the proprietary specification available for adoption by any other company, so it is inherently not interoperable with any products made by other companies. The development process of a proprietary specification is completely closed in that there is no opportunity for collaboration by other companies interested in developing interoperable products. A completely closed development process for a pro-prietary specification involves a single company so issues of procedural fairness, consensus decision making, recording dissent, or dealing with pro-cedural violations are irrelevant. The process has no transparency and there-fore no public oversight. The developer does not publish meeting minutes or provide any record of deliberations, discussions, names of participants, or areas of internal dissent. The participatory and informational aspects of the development process are completely closed.

A proprietary specification is also completely closed in its implementation. The developer will use the specification to ensure interoperability within its own product lines but will not make the specification available, even for a fee. The result is not only a lack of competing, interoperable products but also the inability of governments, interested parties, or the public to view the specification and hold the company accountable for any public policy implications, such as individual privacy, that may exist within the specifica-tion. A significant characteristic of a proprietary specification is that the developer owns all the underlying intellectual property rights and will not necessarily license these rights to another company under any terms.

The result of a completely proprietary specification is that users can become locked into a single company's product lines. Government adoption of closed specifications can be especially problematic. As mentioned earlier, the use of multiple, incompatible proprietary technologies can impede gov-ernment services in critical areas such as public safety, disaster response, and national security. The lack of availability of competing implementations can result in higher costs passed ultimately on to taxpayers. The use of closed formats for services such as the archiving of government documents can force citizens to buy a specific product to access public information, and make the public's ability to access public documents in the future dependent on a single company maintaining the specification and providing backward compatible products. The result of proprietary specifications, as argued in this chapter, is lack of interoperability, single vendor lock-in, lack of com-petition, and government dependence on a single vendor to perform fun-damental government responsibilities.

## A Framework for Open Internet Standards Governance

In contrast to a completely closed specification, a completely open stan-dard is one that is open in its development, open in its implementation, and open in its use. If protocol development were always a purely technical exercise and, once developed, if protocols had no political or economic implications, then the nature of the standards-setting process would be irrelevant. But this account has described how interests and values enter the standards-setting process and how protocols can have significant public interest implications. To confer legitimacy on standards-setting institutions that make design decisions with economic and political consequences, the process must adhere to baseline principles of participatory and informational openness, transparency, well-defined procedures and appeals processes, and accountability.[39]

Participatory openness in standards-development processes requires that institutions allow the participation of any interested party regardless of institutional or corporate affiliation, government backing, credentials, and without requiring membership fees. In reality, the membership requirements and openness of standards-setting institutions vary considerably, with traditional Internet standards bodies like the IETF and W3C providing the greatest openness[40] and small consortia involving a handful of private companies typically the most closed. A completely open process should include well-defined procedures for developing and selecting standards, including norms for publicly recording dissent and a publicly available appeals process for those that disagree with the outcome of the standards process. The process should not allow a single interest or small group to dominate decision making but instead require that any decision obtain broad representative agreement among participants. A completely open process avoids "classes" of membership that limit the decision process to select individuals or limit access to materials based on class. If a standards body does have different membership categories with different voting rights, it should disclose this information, as well as the process required for admittance into these different categories. An open standards body should also have well-defined rules for dealing with procedural violations.

Numerous types of transparency underlie completely open standards, as foreshadowed in the transparency recommendations that chapter 3 outlined. These can be further condensed and divided into transparency in the

39. The following requirements for maximal openness encompass many of the requirements described in the previous section, as well as Eddan Katz and Laura DeNardis, "Best Practices in Internet Standards Governance," White Paper Submission to the Internet Governance Forum, August 2006. Available at http://www.intgovforum.org/Substantive_1st_IGF/BestPracticesforInternetStandardsGovernance.pdf.

40. Gary Malkin, "The Tao of IETF, A Guide for New Attendees of the Internet Engineering Task Force," RFC 1718, November 1994.

development process itself and transparency once a standard is completed. Six types of transparency underlie completely open standards development processes (1) disclosure of membership, if there is a formal membership; (2) disclosure of funding sources if applicable; (3) disclosure of the organizational affiliations of participants; (4) disclosure of process, including the general approach to standards selection (whether based on majoritarian rule, rough consensus, or other approach), appeals procedures, and information about who possesses the ultimate authority in standards decisions; (5) disclosure of a standard's intellectual property rights, if any; and (6) recordation and publication of proceedings, minute meetings, and electronic deliberations. This type of information openness is necessary to allow for the possibility of public oversight. To some, these practices might seem like obvious requirements for standards setting, but in reality, many institutions, including ISO, conduct closed door sessions that are not recorded or ever made available for general public inspection.

Beyond the standards-setting process, the open standard itself, the tool necessary to develop products, should be publicly available. Once finalized, a standard should be made publicly available for two reasons. Public awareness and oversight of a standard's policy implications, as well as technical repercussions, are not possible without the ability to view the actual standard. Standards are blueprints for creating products, not actual software or hardware products themselves. Publishing a standard is not only necessary for public oversight but is the lynchpin of enabling economic competition and therefore maximum innovation of products based on the specification. An unpublished standard is truly a proprietary standard that precludes the possibility of innovation based on the standard. For a standard to be open in its implementation, there must also be no fee associated with accessing the standard. In contrast, many standards-setting institutions charge fees to access and view standards. The most controversial characteristic of maximum openness in a standard's implementation addresses the issue of intellectual property. A maximally open standard is available to implement in products on an irrevocable royalty-free basis. The holder of intellectual property rights should disclose these rights on an ex ante basis. At a minimum, the policies of standards bodies should prohibit rights holders from enforcing patents against the standard's implementation if they fail to disclose these rights during the development and selection process.

Once implemented within a hardware or software product, an open standard has effects on user choice, on competition and innovation, and on technical interoperability. A freely published standard enables multiple companies to develop competing products based on the standard. A standard available without intellectual property restrictions allows companies to develop competing but interoperable products without licensing costs that raise the price of technology products for consumers. The effect is to maximize the possibility of competitive offerings thereby avoiding single vendor lock-in and providing maximum user choice.

This definition of an open standard represents the maximum level of openness based on political, economic, and technical criteria. Whereas economic frameworks for standards use a narrow definition that addresses implications for market competition, innovation, and free trade, the preceding framework suggests a broader range of values to acknowledge the unique public policy role of technical standards.

Considering the enormous variety of information technologies fueling the information society, the appropriate question to consider may not be "open or proprietary" or "how much openness?" but rather "what openness requirements are appropriate to *this* context?" From an economic standpoint that seeks to maximize market competition, the availability of open technical standards may always be desirable for innovation and maximum network effects. The inherent technical interoperability open standards enable may always be critical for technologies involving large-scale communications and information exchange and the associated network effects of these systems. This technical interoperability is not purely an instrumental concern but has broader cultural repercussions to ways in which individuals engage in social interaction, political critique, and technological innovation. Yochai Benkler has suggested that twentieth-century policy choices in the United States, including policies addressing licensing and standards, produced mass industrial media structures promoting passivity and a detached political culture among the general population.[41] In contrast, the more modern open standards approach underlying TCP/IP and other core Internet protocols has produced an architecture promoting greater individual freedom online, the rise of user collaboration, and more active public engagement in technical innovation. Standards with direct bearing on issues of public concern— such as civil liberties online, national defense, the archiving of electronic government documents, and critical infrastructure protection—require a high degree of openness in the standards-setting process.

41. Yochai Benkler, *The Wealth of Networks*, New Haven: Yale University Press, 2006, pp. 176–210.

## The IETF as an Open Institution

Internet standards governance is an example of nongovernmental international rulemaking by entities with significant material interests in the outcome. How to most appropriately legitimate this process has been a central question in Internet governance. This section generally assesses the procedures of the IETF against this framework of Internet standards governance based on principles of openness, transparency, and free market competition. This account of IPv6, overall, has traced the development and implications of a single Internet protocol and thus has focused on the procedures and historical norms of the IETF. The IETF's standards-development process is open in several respects. The IETF has no formal membership requirements and any interested party may freely participate. The IETF's tradition is to consider individual contributors as "individuals" rather than as representatives of corporations, governments, or other organizations, although the majority of individuals receive funding/salaries from their employers and represent the viewpoints of these institutions. The IETF has well-defined procedures for standards ascent, dispute resolution, appeals procedures, and for ongoing support of the standards.

Chapter 2 described how the IETF formalized its commitment to participatory openness, grassroots decision making, and rough consensus after the Kobe affair in which the IAB unilaterally selected a standard to become the next generation Internet protocol. The IETF provides a high degree of transparency and accountability by making meeting minutes, draft standards, electronic discussion forums, and proceedings publicly available via the Internet. Chapter 3 described how this inherent informational openness enabled privacy advocates to become aware of the civil liberties issues under consideration during the design of the IPv6 address structure.

The IETF also meets many criteria of openness in implementation, including freely publishing its standards. The IETF has historically also preferred technologies with no known claims of intellectual property rights, or if the standard has intellectual property claims, the IETF offers royalty-free licensing. Despite this strong preference, the IETF has no official intellectual property requirement that must always be met, other than placing a strong emphasis on upfront disclosure of intellectual property rights. The W3C's royalty-free policy provides a higher degree of openness in this regard. As mentioned, the Internet's underlying protocols have historically been available on a predominantly royalty-free basis, a characteristic contributive to the Internet's rapid growth, product innovations, and democratic participation. But it is also evident that mandating the use of royalty-free standards can produce inadvertent consequences such as eliminating the possibility of using popular royalty-bearing Wi-Fi and GSM standards.

The IETF's openness conveys the impression that Internet standards development, generally, is open. As described earlier, other standards bodies develop many of the information technology standards necessary to enable information exchange over the Internet. Many of these institutions are much more closed than the IETF. For example, ISO exhibits a relatively closed standards-development approach. ISO is an international standards-setting organization comprised of national standards bodies from more than 150 countries. As another example, the American National Standards Institute (ANSI) serves as the standards organization representing the United States in ISO. Membership is limited to the extent that only national standards institutions "most representative of standardization in their country (one member in each country) can join."[42] These member organizations each pay a fee to participate and individuals are not eligible for direct membership unless participating through a national standards institution that is an ISO member. Another significant contrast is that ISO does not make working documents publicly available, instead considers these works in progress as internal documents. Once standards are finalized, ISO charges a fee for accessing the document rather than making specifications freely available.

The ITU also sets many Internet-related standards in areas ranging from traditional telecommunications to optical transmission systems to Internet Protocol Television (IPTV). The ITU, in 2007, decided to make its standards (called "recommendations") freely available to the public. However, membership is not open commensurate with IETF norms. The ITU provides some avenues for anyone to participate in standards development workshops but only opens official membership to national government members of the United Nations and private sector members who pay significant annual fees.

On a spectrum of possible openness, the IETF's degree of procedural, participatory, and informational openness is high. Michael Froomkin has described the IETF's Internet standards process as a case study in Habermasian discourse ethics. Froomkin notes the striking similarities between the IETF's standards-development procedures and Habermas's "account of the properties that a practical discourse requires in order to legitimate the

---

42. See ISO membership information. Accessed at http://www.iso.org/iso/about/ discover-iso_meet-iso/discover-iso_who-can-join-iso.htm.

rules it produces."[43] Froomkin's analogy is appropriate in several respects. Habermasian discourse requires that all relevant voices should get a hearing, as all voices can potentially contribute to IETF discourses. It also suggests that the best communicative approaches require the presentation of the best arguments available at the present time. The IETF's philosophy of using proven "running code" seems to meet this criterion. Its "rough consensus" philosophy also fits within the Habersian vision of ideal discourse. The IETF's participatory openness allows for the *potential* for all voices to be heard, unlike other standards processes, and this is a critical component in legitimating the rules it produces.

Because technological protocols have direct social implications, it is appropriate to view the standards-setting process as part of the broader public sphere and fitting to draw similarities between attempts at ideal discourses and the IETF's processes. But this account has also provided caveats about the ways in which technical standards-setting inherently falls short of these idealized views. This account has discussed some intrinsic barriers to participation related to technical expertise, language, funding, and culture. It has further examined the role of competing institutional tensions and corporate rivalries in standards development. It has described ways in which developing countries, as later entrants, can be left out of the standards-setting process. It has described how standards create scarce resources that become a struggle for control over global information architectures. Once developed, standards can become a form of technological discourse used to advance various ideologies or political and economic objectives. Finally, the mere development of a standard does not translate into the implementation of a standard, raising a question about the intrinsic worth of an open standard until it is actually translated into implementation and use, a question addressed in the following section.

## The Limits of Technical Inevitability

The history of the depletion of the Internet address space and the slow deployment of IPv6 clearly demonstrates the limitations of openness in standardization as well as the limitations of direct government intervention in Internet standards adoption. The design and availability of an open standard does not necessarily translate into its implementation, and

43. A. Michael Froomkin, "Habermas@discourse.net: Toward a Critical Theory of Cyberspace," 116 *Harvard Law Review* 749–873 (January 2003).

government mandates do not, nor should they, automatically trigger widespread adoption.

It is a mathematical reality that the store of IPv4 addresses has diminished to the point of becoming critically scarce. In 2007, an IETF statement warned that the "IPv4 free address pool will be exhausted within no more than 3–4 years, and possibly sooner. At that point it will become increasingly difficult for ISPs and end sites to obtain the public IPv4 address space they need to expand operations."[44] Statements by ICANN and all five regional Internet registries have similarly issued warnings about the imminent exhaustion of the IPv4 address space. These projections have assumed that Internet address allocations and assignments would continue at a consistent pace relative to historical allocations. Predicting when the supply of IPv4 addresses will, in reality, "go to zero" is not an exact science. Recall that predictions that the Internet address space would become critically scarce began in the early 1990s. At a Vancouver meeting of Internet engineers in 1990, some participants forecasted that at the current address assignment rate, the Internet address space would be depleted by 1994. The European Union forecasted that IPv4 addresses would become "critically scarce" by 2005, again a time frame elapsing without any catastrophic Internet collapse. In subsequent years, predicted target dates for address depletion have come and gone and have been consistently incorrect. But without significant technological or political intervention, it is safe to state that "at some point in the future," the IPv4 address space will eventually be completely depleted.

IPv6 was designed to solve this problem of Internet address space exhaustion. The core IPv6 standards were completed long ago and IPv6 is widely available in products. But IPv6 adoption has been anemic. More than a decade after the publication of the IPv6 specification, the migration to IPv6 simply had not occurred. Predictions of imminent IPv6 migration have shadowed predictions of imminent IPv4 address exhaustion. National government mandates called for upgrades by 2005, and Internet engineers themselves expected that the world would widely implement IPv6 long before the IPv4 address space would become critically scarce.

As the IETF has succinctly described the real world situation, "widespread deployment has barely begun."[45] The IP address space has continued to

44. Tom Narten, "IETF Statement on IPv4 Exhaustion and IPv6 Deployment," Informational Internet Draft, November 12, 2007.
45. Ibid.

decline, IPv6 advocacy groups have enthusiastically pushed IPv6 migration, vendors have incorporated the new standard into products, and governments have mandated adoption. Yet years have elapsed with no significant IPv6 use across the global Internet. Like the history of the metric system standard, the history of IPv6 demonstrates the social construction of notions of technical inevitability and technical resistance. The centuries-long American rejection of the metric system reinforces how standards are social conventions and portends that upgrading to a new standard in the face of perceived international inertia is not preordained.

Protocol adoption is where markets meet protocols. At its inception, engineers selected IPv6 outside of the mechanisms of market economics, identifying the need for a new protocol because of anticipated Internet address scarcity, rather than basing their decisions on the views of large information technology users who already had sufficient addresses. The distribution of IP addresses has also occurred extraneous to markets, with addresses never bought and sold in traditional exchanges. But dominant market forces clearly enter the realities of protocol adoption.

Even if IPv6 never gains considerable momentum, it has influenced the Internet's technical, institutional, and legal architecture in concrete ways. First, institutional struggles the Internet standards community faced during the selection of the next generation Internet protocol led to the articulation and formalization of its grassroots and open approach to standards setting, including David Clark's famous statement, "We reject: kings, presidents and voting. We believe in rough consensus and running code." Second, in the uncertain context of whether OSI protocols would replace the TCP/IP protocols underlying the Internet, the selection of IPv6 essentially rejected an OSI alternative and elevated the status of TCP/IP. Third, IPv6 selection reinforced the power of the IETF as the controlling institution over the core Internet protocols rather than relinquishing change control to another standards organization. Fourth, design decisions faced during the development of IPv6 also reinforced institutional norms about architecting privacy considerations into Internet protocols. One question that IPv6 has not resolved is the ongoing conflict over who should have centralized oversight of the Internet address space.

When Internet registries assign the last IPv4 addresses, the Internet will continue to operate. However, those without an existing store of addresses will be at a distinct disadvantage and will likely have to implement IPv6 and some sort of transition mechanism (e.g., NAT-PT). One complication is that most transition approaches require IPv4 addresses. Even NAT-PT requires a small number of shared public IPv4 addresses. The depletion of the address space, coupled with the unexpectedly slow deployment of IPv6, could have many social, political, and institutional implications and could potentially shape the nature of the Internet's technical architecture. The following sections describe these possible repercussions.

### Heightened Global Inequity

Years ago, Vinton Cerf suggested, "The value of IPv6 can be realized only if the deployment effort is broadly based on a global scale."[46] Without economic incentives or a new "killer" application requiring IPv6, it is unlikely that the existing global base of Internet users will upgrade to IPv6 en masse. Those in parts of the world without large existing stores of IPv4 addresses will, worst case scenario, potentially experience access shortages. More likely, new users or those requiring additional capacity will upgrade to IPv6, an upgrade that carries a set of network management and security complexities and that, without the accompaniment of transitional mechanism, could hinder user access to existing IPv4 web servers and other Internet resources. Organizations, individuals, or service providers supporting both IPv6 and IPv4 could face additional challenges related to network management and security complexities and greater resource requirements (both technical and intellectual). If Internet address exchange markets proceed, these market-based systems could potentially increase global inequities in access to knowledge because those who can afford to pay premiums for global Internet addresses will benefit relative to emerging markets, already with a smaller share of Internet addresses.

### Internet Governance Conflicts

Global tensions over control of critical Internet resources have existed for years, but increased depletion of the IPv4 address space will only fuel tensions among the institutions, nations, and intergovernmental organizations wanting to either preserve or increase oversight of these resources. The complete exhaustion of the Internet address space without an accompanying transition to IPv6 could result in greater government attempts to intervene in Internet technical architecture, possibly resulting in heightened transnational jurisdictional conflicts and more highly regulated technical architectures in some regions. A philosophical shift in Internet governance that starts viewing Internet addresses as resources to be bought and sold in free markets rather than as common public resources will also

---

46. Vinton Cerf. Quoted on opening web page of European IPv6 Task Force. Accessed at http://www.eu.ipv6tf.org/in/i-index.php on February 16, 2005.

have implications. On the surface, the idea of address markets has pragmatic appeal, even if only a stopgap, temporary measure. In practice, introducing address exchange markets could have unintended consequences such as attracting stricter government regulation of standards and of the Internet address space. It might also induce a permanent transformation: if the world begins exchanging IPv4 addresses in free markets, what would be the rationale for not exchanging IPv6 addresses in a similar fashion? If history has provided any lessons about Internet addresses, it suggests that with the pace of innovation and the sudden emergence of new applications, it is conceivable that the IPv6 address space could one day become scarce.

### Architectural and Political Restructuring

Most significantly, IPv4 address depletion and the gradual introduction of IPv6 could architecturally transform parts of the Internet. As this book has argued, architectural changes are also changes in arrangements of power. It is conceivable that segments of the Internet could balkanize into separate IPv4 and IPv6 islands. IPv6-only regions would likely connect to the prevailing IPv4 Internet via translation, but only by adding technical complexity and security and access challenges. One can easily imagine these translation points as control points facilitating filtering, surveillance, and censorship by repressive governments seeking to selectively block access. On a more technical level, some Internet applications might not perform well through widespread translation devices. Protocol fragmentation would not be problematic in all cases. IPv6-only networks could support applications not requiring global reach and universal interoperability, such as regional control networks for water and energy systems, or for specific surveillance and monitoring systems (e.g., sensors) that inherently address closed systems.

Another implication is the further reversal of the already eroded end-to-end architectural principle that influenced the original design of Internet protocols and that contributed to the resulting decentralization of information production and innovation. The prevalence of network firewalls and NAT devices has already contravened this principle, but more widespread structural deployment of translation devices could make this principle irrelevant, as well as complicating end-to-end security and performance management. One uncertainty is how this further disruption of the end-to-end principle will affect the universality and relative openness of the Internet as well as its ability to support decentralized innovation and information production and promote the global flow of knowledge over the Internet.

The history of unexpected Internet developments points to the possibility of a radically new application that either encourages global IPv6 adoptions, supersedes the problem of address scarcity, or calls for the development of a completely new Internet protocol. Regardless of outcome, this form of protocol politics, operating at levels of abstraction and within institutional structures outside the bounds of traditional governance, will continue to emerge as the transnational rulemaking structure of the global information society.

# List of Abbreviations

| | |
|---|---|
| 3G | Third generation wireless |
| AD | Area director (within IETF) |
| ADSL | Asymmetric Digital Subscriber Line |
| AfriNIC | African Network Information Centre |
| ANSI | American National Standards Institute |
| APNIC | Asia Pacific Network Information Centre |
| ARIN | American Registry for Internet Numbers |
| ARP | Address Resolution Protocol |
| ARPA | Advanced Research Projects Agency |
| ARPANET | Advanced Research Projects Agency Network |
| ATM | Asynchronous Transfer Mode |
| BBN | Bolt Beranek and Newman |
| B-ISDN | Broadband Integrated Services Digital Network |
| BIT | Binary digit |
| BOF | Birds of a Feather Group (within IETF) |
| BSD | Berkeley Software Distribution |
| CAS | Chinese Academy of Science |
| CATNIP | Common Architecture for the Internet |
| CCNIC | China Internet Network Information Center |
| ccTLD | Country code top-level domain |
| CDPD | Cellular Digital Packet Data |
| CDT | Center for Democracy and Technology |
| CERNET | China Education and Research Network |
| CERT | Computer Emergency Response Team |

| | |
|---|---|
| CIDR | Classless interdomain routing |
| CIO | Chief information officer |
| CLNP | ConnectionLess Network Protocol |
| CNGI | China Next Generation Internet |
| CNRI | Corporation for National Research Initiatives |
| DARPA | Defense Advanced Research Projects Agency |
| DBMS | Database management system |
| DCOS | Dynamic Coalition on Open Standards |
| DDN-NIC | Defense Data Network-Network Information Center |
| DDoS | Distributed denial of service |
| DEC | Digital Equipment Corporation |
| DHCP | Dynamic Host Configuration Protocol |
| DMCA | Digital Millennium Copyright Act |
| DNS | Domain name system |
| DNSSEC | Domain Name System Security Extensions |
| DoD | Department of Defense |
| DRM | Digital Rights Management |
| DSL | Digital subscriber line |
| EFF | Electronic Frontier Foundation |
| EPIC | Electronic Privacy Information Center |
| EPRI | Electric Power Research Institute |
| EU | European Union |
| FEMA | Federal Emergency Management Agency |
| F/LOSS | Free/Libre Open Source Software |
| FNC | Federal Networking Council |
| FTP | File Transfer Protocol |
| GAO | Government Accountability Office |
| GDP | Gross Domestic Product |
| GIG | Global Information Grid |
| GOSIP | Government Open Systems Interconnection Protocol |
| GPRS | General Packet Radio Service |
| GSM | Global System for Mobile communications |

| | |
|---|---|
| HCI | Human-computer interaction |
| HP | Hewlett Packard |
| HTML | HyperText Markup Language |
| HTTP | HyperText Transfer Protocol |
| IAB | Internet Architecture Board or Internet Activities Board |
| IANA | Internet Assigned Numbers Authority |
| IBM | International Business Machines |
| ICANN | Internet Corporation for Assigned Names and Numbers |
| ICCB | Internet Configuration Control Board |
| ICT | Information and communication technologies |
| IEC | International Electrotechnical Commission |
| IEEE | Institute of Electrical and Electronics Engineers |
| IESG | Internet Engineering Steering Group |
| IETF | Internet Engineering Task Force |
| IGF | Internet Governance Forum |
| IGP | Internet Governance Project |
| IMP | Interface Message Processor |
| IP | Internet Protocol |
| IPAE | Internet Protocol Address Encapsulation |
| IPng | Internet Protocol next generation |
| IPR | Intellectual property rights |
| IPsec | Internet Protocol Security |
| IPv4 | Internet Protocol version 4 |
| IPv6 | Internet Protocol version 6 |
| IPTV | Internet Protocol Television |
| IPX/SPX | Internetwork Packet Exchange/Sequenced Packet Exchange |
| IRTF | Internet Research Task Force |
| ISO | International Organization for Standardization |
| ISOC | Internet Society |
| ISP | Internet service provider |
| ITARS | International Traffic in Arms Regulations |
| ITU | International Telecommunication Union |

| | |
|---|---|
| ITU-T | ITU Telecommunications Sector |
| JPEG | Joint Photographic Experts Group |
| LACNIC | Latin America and Caribbean Network Information Centre |
| LAN | Local area network |
| LIR | Local Internet registries |
| MAC | Media access control |
| MP3 | MPEG-1 Audio Layer 3 |
| MPEG | Motion Picture Experts Group |
| NAT | Network address translation |
| NAT-PT | Network address translation-protocol translation |
| NAv6TF | North American IPv6 Task Force |
| NEMO | Network Mobility Basic Support Protocol |
| NIC | Network interface controller |
| NII | National Information Infrastructure |
| NIR | National Internet registries |
| NIST | National Institute of Standards and Technology |
| NRO | Number Resource Organization |
| NSAP | Network service access point |
| NSFNET | National Science Foundation Network |
| NSI | Network Solutions |
| NTB | Nontariff barriers |
| NTIA | National Telecommunications and Information Administration |
| NWG | Network Working Group |
| OASIS | Organization for the Advancement of Structured Information Standards |
| ODF | Open Document Format |
| OMB | US Office of Management and Budget |
| OOXML | Open Office XML |
| OPES | Open Pluggable Edge Services |
| OSI | Open Systems Interconnection |
| OUI | Organizationally unique identifier |
| P3P | Platform for Privacy Preferences |

| | |
|---|---|
| PIP | "P" Internet Protocol |
| POISED | Process for Organization of Internet Standards (working group) |
| RAND | Reasonable and nondiscriminatory |
| RARE | Reseaux Associés pour la Recherche Européenne |
| RFC | Request for comments |
| RIPE NCC | Réseaux IP Européens-Network Coordination Centre |
| RIR | Regional Internet registry |
| ROAD | ROuting and ADdressing |
| SIP | Simple Internet Protocol |
| SIPP | Simple Internet Protocol Plus |
| SMDS | Switched Multimegabit Data Service |
| SMTP | Simple Mail Transfer Protocol |
| SNA | Systems Network Architecture |
| SONET | Synchronous Optical Network |
| SRI NIC | Stanford Research Institute's Network Information Center |
| STS | Science and technology studies |
| TBT | Technical barriers to trade |
| TCP | Transmission Control Protocol |
| TCP/IP | Transmission Control Protocol/Internet Protocol |
| TLD | Top-level domain |
| TUBA | TCP and UDP with Bigger Addresses |
| UDP | User Datagram Protocol |
| UDRP | Uniform Domain-Name Dispute-Resolution Policy |
| UN | United Nations |
| URL | Uniform resource locator |
| US-CERT | United States Computer Emergency Readiness Team |
| USC-ISI | University of Southern California Information Sciences Institute |
| VoIP | Voice over Internet Protocol |
| W3C | World Wide Web Consortium |
| WAPI | Wireless Local Area Network Authentication and Privacy Infrastructure |

| WCA | Wireless Communication Association |
| WGIG | Working Group on Internet Governance |
| Wi-Fi | Wirelesss Fidelity |
| WiMAX | Worldwide Interoperability for Microwave Access |
| WIPO | World Intellectual Property Organization |
| WSIS | World Summit on the Information Society |
| WTO | World Trade Organization |
| XML | eXtensible Markup Language |

# Technical Appendix

For those interested, the following sections explain shorthand notation for IPv4 and IPv6 addresses and the frame header formats for IPv4 and IPv6.

## Shorthand Notation for IPv4 and IPv6 Addresses

### IPv4

This technical appendix describes how the 32-bit IP address translates into dotted decimal format. The shorthand convention involves a conversion from the binary numbering system (using two digits) that computers understand to the decimal numbering system (using ten digits) that humans use in real life.

The mathematical conversion between the computer-readable 32-bit address and the human-readable dotted decimal format address includes three steps: dividing the 32-bit address into four octets (groups of 8 bits), converting each octet into its decimal equivalent, and placing "dots" between each of the four derived decimal numbers. The following is an example of this conversion:

**Computer Readable IP Address: 00011110000101011100001111011101**
Divide the IP address into four octets (groups of 8 bits):

00011110

00010101

11000011

11011101

Convert each binary octet into its equivalent decimal number:

$00011110 = 16 + 8 + 4 + 2 = 30$

$00010101 = 16 + 4 + 1 = 21$

$11000011 = 128 + 64 + 2 + 1 = 195$

$11011101 = 128 + 64 + 16 + 8 + 4 + 1 = 221$

Write out the decimal values separated by dots:

**Human Readable IP Address: 30.21.195.221**    This "dotted decimal format" is much easier for humans to comprehend, discuss, track, and manage but not useful to networking equipment.

### IPv6

Shorthand notation is even more important for 128-bit IPv6 addresses.[1] The following is an IPv6 address:

0111010010011101100001101010101110111101000110010011001001001
0011101001001110110000110101011101111101000110010011001001001001
1010111000

Just like "dotted decimal format" is used as shorthand for an IPv4 address, IPv6 has its own shorthand representation:

X:X:X:X:X:X:X:X

where each X is equal to the hexadecimal representation of 16 bits. The convention for IPv6 notation is to use the hexadecimal numbering system. The following is a random example of an IPv6 address in shorthand notation:

FDDC:AC10:8132:BA32:4F12:1070:DD13:6921

Note that the above shorthand representation of an IPv6 address consists of eight groups of four hexadecimal numbers separated by colons. Each hexadecimal number represents four binary numbers as follows:

| Hexadecimal numeral | Binary equivalent |
|---|---|
| 0 | 0000 |
| 1 | 0001 |
| 2 | 0010 |
| 3 | 0011 |
| 4 | 0100 |

1. The conventions for IPv6 Notation appear in Robert Hinden and Steve Deering, "Internet Protocol Version 6 (IPv6) Addressing Architecture," RFC 3513, April 2003.

| | |
|---|---|
| 5 | 0101 |
| 6 | 0110 |
| 7 | 0111 |
| 8 | 1000 |
| 9 | 1001 |
| A | 1010 |
| B | 1011 |
| C | 1100 |
| D | 1101 |
| E | 1110 |
| F | 1111 |

Therefore the shorthand representation of an IPv6 address can be translated as follows:

FDDC = 1111110111011100
AC10 = 1010110000010000
8132 = 1000000100110010
BA32 = 1011101000110010
4F12 = 0100111100010010
1070 = 0001000001110000
DD13 = 1101110100010011
6921 = 0110100100100001

Putting it all together, the "human readable" address:

FDDC:AC10:8132:BA32:4F12:1070:DD13:6921

is equivalent to the actual "machine readable" IPv6 address:

1111 1101 1101 1100 1010 1100 0001 0000 1000 0001 0011 0010 1011 1010
0011 0010 0100 1111 0001 0010 0001 0000 0111 0000 1101 1101 0001
0011 0110 1001 0010 0001

As cumbersome as the hexadecimal version appears, it is a considerable improvement over writing out the entire 128-bit string of 0s and 1s as above.

**Compressing the Address Further: X:X::X:X**    Many IPv6 addresses contain long strings of 0s, and notation conventions can further compress these addresses. For example, the hexadecimal representation

ADFD:0000:0000:0000:0000:0000:1357:3A11

is customarily shortened to

ADFD:0:0:0:0:0:1357:3A11

by dropping the "leading zeros" in each group. To compress this even further, the symbol "::" indicates one or more groups of 16 bits of zeros:

ADFD::1357:3A11

Sometimes an older IPv4 address is incorporated into an IPv6 address. The framework for this notation is

X:X:X:X:X:X:d.d.d.d

where the Xs are hexadecimal representations of 16-bit groups and the ds represent standard dotted decimal format. An example of this notation is the following:

0:0:0:0:0:FFFF:15.129.55.9

**Header Frame Formats for IPv4 and IPv6**

Each packet of information traversing the Internet contains not only information (payload) but a header providing administrative and routing information about the packet. The header contains the source and destination address, for example. The following are the header formats for IPv4 and IPv6. Note that the IPv6 header format is significantly simplified relative to the IPv4 header format.

**IPv4 Header Format[2]**

| Vers. | IHL | Type of service | | Total Length | |
|---|---|---|---|---|---|
| Identification | | | Flags | Fragment Offset | |
| TTL | | Protocol | Header Checksum | | |
| Source Address | | | | | |
| Destination Address | | | | | |
| Options | | | | | Padding |

Version:                4-bit Internet Protocol version number = 4
IHL:                    4-bit Internet header length

2. Jon Postel, editor. "Internet Protocol: DARPA Internet Program Protocol Specification," RFC 791, September 1981.

Type of service:        8 bits specifying precedence of information
Total length:           16 bits, total length of datagram in octets
Identification:         A sender assigned value to aid fragment assembling
Flags:                  3-bit control flag such as "last fragment"
Fragment offset:        13 bits indicating where fragment belongs in datagram
TTL:                    8-bit time to live
Protocol:               8-bit identification of next level protocol
Header checksum:        16-bit error detection procedure
Source address:         32-bit source Internet address
Destination address:    32-bit destination Internet address
Options:                Variable length field for optional information
Padding:                Variable length superfluous bits ensuring header ends on 32-bit boundary

## IPv6 Header Format[3]

| Version | Traffic Class | Flow Label | | |
|---|---|---|---|---|
| Payload Length | | | Next Header | Hop Limit |
| Source Address | | | | |
| Destination Address | | | | |

Version:                4-bit Internet Protocol version number = 6
Traffic class:          8-bit traffic class field
Flow label:             20-bit flow label
Payload length:         16-bit assigned integer specifying IPv6 payload length
Next header:            8-bit selector identifying type of header following IPv6 header
Hop limit:              8-bit integer decremented by 1 for each node forwarding the packet. Packet is discarded if hop limit is decremented to zero.
Source address:         128-bit address of packet originator
Destination address:    128-bit address of intended packet recipient

3. Steven Deering and Robert Hinden, "Internet Protocol, Version 6 (IPv6) Specification," RFC 2460, December 1998.

# Selected Bibliography

The following list of selected texts represents a partial record of the works and sources consulted in researching and formulating *Protocol Politics*.

This book could not have progressed without the immense historical archive of documents, RFCs, meeting minutes, and Internet mailing lists chronicling personal conversations and debates within institutions directly involved in Internet standardization issues. Mailing lists provide the arena for debates and are the mechanism for participation in Internet standards setting. These first-person postings provided a snapshot of what participants expressed in situ rather than through retrospective accounts. Many of these mailing lists existed prior to the development of the web and prior to widespread public access to the Internet. The mailing list archives that contributed the most to this research chronicled dialogues with individuals directly involved in protocol selection, design, and implementation processes, among them key participants in standards institutions including the IETF, the IAB, and the IESG. Countless IETF working group lists were available at the IETF website. Some useful mailing lists also included info.big-internet, info.ietf, and comp.protocols.tcp-ip. IETF working group documents provided considerable technical information about IPv6. The "IPv6 working group," formerly the "IPng working group," provided the most useful technical resource for IPv6 specifications. Archives of meeting minutes and discussions from 1994 through the present document the deliberations of the IPng working group, the SIPP working group, the IPv6 transition working group, and the renamed IPv6 working group.

IPv6 is not a single specification. IPv6 implementations engage countless technical specifications depending on requirements, such as operating IPv6 over various types of networks, IPv6 management, and information compression over IPv6. An electronic archive accessed at http://playground. sun.com/pub/ipng/html/ipng-main.html provided more than fifty of the

IPv6-related technical specifications, along with working group minutes and a chronicle of IPv6 product implementations within various operating systems and routers. Additionally, informational RFCs and standards-track RFCs were accessed at http://www.rfc-editor.org/

The IETF website, www.ietf.org, also archives more than 40,000 pages of proceedings from its triennial conferences held since the institution's 1986 inception. The minutes of the monthly IAB meetings also provided a snapshot of debates about IPv6 among participants in the Internet's technical community.

Abbate, Janet. *Inventing the Internet.* Cambridge: MIT Press, 1999.

Aitken, Hugh G. J. "Allocating the Spectrum: The Origins of Radio Regulation." 35 *Technology and Culture* 686–716 (October, 1994).

Aitken, Hugh G. J. *The Continuous Wave: Technology and American Radio, 1900–1932.* Princeton: Princeton University Press, 1985.

Alder, Ken. *The Measure of All Things: The Seven-Year Odyssey and Hidden Error that Transformed the World.* New York: Free Press, 2002.

Alder, Ken. "A Revolution to Measure: The Political Economy of the Metric System in France." In *Values of Precision*, M. Norton Wise, ed. Princeton: Princeton University Press, 1995, pp. 39–71.

Balkin, Jack. "Digital Speech and Democratic Culture: A Theory of Freedom of Expression for the Information Society." Yale Law School, Public Law Working Paper 63. 79 *New York University Law Review* 1–58 (2004).

Benkler, Yochai. "Coase's Penguin, or, Linux and The Nature of the Firm." 112 *Yale Law Journal* 369–446 (Winter 2002–03).

Benkler, Yochai. *The Wealth of Networks: How Social Production Transforms Markets and Freedom.* New Haven: Yale University Press, 2006.

Benoliel, Daniel. "Technological Standards, Inc.: Rethinking Cyberspace Regulatory Epistemology." 92 *California Law Review* 1069 (2004).

Berners-Lee, Tim, with Mark Fischetti. *Weaving the Web: The Original Design and Ultimate Destiny of the World Wide Web by Its Inventor.* New York: HarperCollins, 1999.

Bijker, Wiebe. *Of Bicycles, Bakelites, and Bulbs: Toward a Theory of Sociotechnical Change.* Cambridge: MIT Press, 1995.

Bijker, Wiebe E., Thomas P. Hughes, and Trevor Pinch. *The Social Construction of Technological Systems: New Directions in the Sociology and History of Technology.* Cambridge: MIT Press, 1999.

Blumenthal, Marjory S., and David D. Clark. "Rethinking the Design of the Internet: The End-to-End Arguments vs. the Brave New World." In *Communications Policy in Transition: The Internet and Beyond*, Benjamin M. Compaine and Shane Greenstein, eds. Cambridge: MIT Press, 2001, pp. 70–109.

Borgmann, Albert. *Technology and the Character of Contemporary Life: A Philosophical Inquiry*. Chicago: The University of Chicago Press, 1984.

Bradner, Scott, and Allison Mankin, eds. *IPng: Internet Protocol Next Generation*. North Reading, MA: Addison-Wesley, 1996.

Bradner, Scott, and Allison Mankin, eds. "IP: Next Generation (IPng) White Paper Solicitation." RFC 1550. December 1993.

Bradner, Scott, and Allison Mankin. "The Recommendation for the IP Next Generation Protocol." RFC 1752. January 1995.

Britton, Edward, and John Tavs. "IPng Requirements of Large Corporate Networks." RFC 1678. August 1994.

Bush, Randy, and David Meyer. "Some Internet Architectural Guidelines and Philosophy." RFC 3439. December 2002.

Callon, Ross. "TCP and UDP with Bigger Addresses (TUBA), A Simple Proposal for Internet Addressing and Routing." RFC 1347. June 1992.

Carlson, Richard, and Domenic Ficarella. "Six Virtual Inches to the Left: The Problem with IPng." RFC 1705. October 1994.

Carpenter, Brian. "Architectural Principles of the Internet." RFC 1958. June 1996.

Carpenter, Brian. "Middleboxes: Taxonomy and Issue." RFC 3234. February 2002.

Carpenter, Brian. "IPng White Paper on Transition and Other Considerations." RFC 1671. August 1994.

Castells, Manuel. *The Internet Galaxy: Reflections on the Internet, Business, and Society*. Oxford: Oxford University Press, 2001.

Cerf, Vinton. "I Remember IANA." RFC 2468. October 1998.

Cerf, Vinton. IAB Recommended Policy on Distributing Internet Identifier Assignment and IAB Recommended Policy Change to Internet "Connected Status." RFC 1174. August 1990.

Cerf, Vinton. "The Internet Activities Board." RFC 1160. May 1990.

Ceruzzi, Paul. *A History of Modern Computing*, 2nd ed. Cambridge: MIT Press, 2003.

Chander, Anupam. "The New, New Property." 81 *Texas Law Review* 715–98 (February 2003).

Chandler, Alfred D., Jr. *The Visible Hand: The Managerial Revolution in American Business*. Cambridge, MA: Belknap Press, 1977.

Clark, David. "The Design Philosophy of the DARPA Internet Protocols." 18 *Proceedings of SIGCOMM 88, ACM COR* 106–14 (August 1988).

Clark, David, et al. "Towards the Future Internet Architecture." RFC 1287. December 1991.

Cohen, Danny. "Working with Jon, Tribute delivered at UCLA, October 30, 1998." RFC 2441. November 1998.

Cranor, Lorrie Faith. "The Role of Privacy Advocates and Data Protection Authorities in the Design and Deployment of the Platform for Privacy Preferences." *Proceedings of the Computers, Freedom, and Privacy Conference*, 2002. Accessed at http://www.cfp2002.org/proceedings/proceedings/cranor.pdf.

Crocker, Steve. "Host Software." RFC 1. April 1969.

Crocker, Steve. "The Process for Organization of Internet Standards Working Group." RFC 1640. June 1994.

Curran, John. "Market Viability as an IPng Criteria." RFC 1669. August 1994.

David, Paul A., and Partha Dasgupta. "Toward a New Economics of Science." 23 *Research Policy* 487–521 (1994).

Davidson, Alan, John Morris, and Robert Courtney. "Strangers in a Strange Land: Public Interest Advocacy and Internet Standards." Center for Democracy and Technology paper, 2002. Accessed at http://tprc.org/papers/2002/97/Strangers_CDT_to_TPRC.pdf.

Deering, Stephen, and Robert Hinden. "Internet Protocol, Version 6 Specification." RFC 1883. December 1995.

Deering, Stephen, and Robert Hinden. "Internet Protocol, Version 6 Specification." RFC 2460. December 1998.

Deering, Stephen, and Robert Hinden. "Internet Protocol, Version 6 (IPv6) Architecture." RFC 2401. December 1998.

Deering, Stephen, and Robert Hinden. "Statement on IPv6 Address Privacy," November 6, 1999. Accessed at http://playground.sun.com/ipv6/specs/ipv6-address-privacy.html.

Dixon, Tim. "Comparison of Proposals for Next Version of IP." RFC 1454. May 1993.

Douglas, Susan J. *Inventing American Broadcasting: 1899–1922*. Baltimore: Johns Hopkins University Press, 1987.

Edwards, Paul N. The *Closed World: Computers and the Politics of Discourse in Cold War America*. Cambridge: MIT Press, 1996.

Egevang, Kjeld, and Paul Francis. "The IP Network Address Translator." RFC 1631. May 1994.

Ellul, Jacque. *The Technological Society.* New York: Vintage Books, 1964.

Epstein, Steven. *Impure Science: AIDS, Activism, and the Politics of Knowledge.* Berkeley: University of California Press, 1996.

Escobar, Arturo. *Encountering Development: The Making and Unmaking of the Third World.* Princeton: Princeton University Press, 1995.

Ezrahi, Yaron. *The Descent of Icarus: Science and the Transformation of Contemporary Democracy.* Cambridge: Harvard University Press, 1990.

Feenberg, Andrew. *Questioning Technology.* London: Routledge, 1999.

Flanagan, Mary, Daniel Howe, and Helen Nissenbaum. "Values in Design: Theory and Practice." In *Information Technology and Moral Philosophy.* Draft, 2005. Accessed at http://www.nyu.edu/projects/nissenbaum/papers/Nissenbaum-VID.4-25.pdf.

Fleck, Ludwik. *Genesis and Development of a Scientific Fact.* Chicago: University of Chicago Press, 1935.

Fleischman, Eric. "A Large Corporate User's View of IPng." RFC 1687. August 1994.

Foucault, Michel. *The Order of Things: An Archaeology of the Human Sciences.* New York: Pantheon Books, 1971.

Froomkin, Michael A. "Habermas@discourse.net: Toward a Critical Theory of Cyberspace." 116 *Harvard Law Review* 749–873 (January 2003).

Fuller, Vince, et al. "Classless Inter-Domain Routing (CIDR): An Address Assignment and Aggregation Strategy." RFC 1519. September 1993.

Galloway, Alexander. *Protocol: How Control Exists after Decentralization.* Cambridge: MIT Press, 2004.

Garnham, Nicholas. "Information Society as Theory or Ideology: A Critical Perspective on Technology, Education, and Employment in the Information Age." 3 *Information, Communication, and Society* 139–52 (2000).

Ghosh, Rishab. *An Economic Basis for Open Standards,* December 2005. *Accessed at* http://flosspols.org/deliverables/FLOSSPOLS-D04-openstandards-v6.pdf.

Gieryn, Thomas. "Boundary-Work and the Demarcation of Science from non-Science: Strains and Interests in Professional Ideologies of Scientists." 48 *American Sociological Review* 781–95 (1983).

Grewal, David Singh. *Network Power: The Social Dynamics of Globalization.* New Haven: Yale University Press, 2008.

Gross, Phill. "A Direction for IPng." RFC 1719. December 1994.

Gross, Phill, and Philip Almquist. "IESG Deliberations on Routing and Addressing." RFC 1380. November 1992.

Hafner, Katie, and Matthew Lyon. *Where Wizards Stay up Late: The Origins of the Internet.* New York: Simon and Schuster, 1996.

Hain, Tony. "Architectural Implications of NAT." RFC 2993. November 2000.

Haraway, Donna J. *Simians, Cyborgs, and Women: The Reinvention of Nature.* New York: Routledge, 1991.

Harding, Sandra. *Whose Science? Whose Knowledge? Thinking from Women's Lives.* Ithaca: Cornell University Press, 1991.

Harris, Susan, ed. "The Tao of IETF—A Novice's Guide to the Internet." RFC 3160. August 2001.

Harvey, David. *The Condition of Postmodernity: An Inquiry into the Origins of Social Change.* Oxford: Blackwell, 1989.

Heidegger, Martin. *The Question Concerning Technology.* New York: Harper Torch-books, 1977.

Hickman, Larry A. *Philosophical Tools for Technological Culture: Putting Pragmatism to Work.* Bloomington: Indiana University Press, 2001.

Hinden, Robert M. "Applicability Statement for the Implementation of Classless Inter-domain Routing (CIDR)." RFC 1517. September 1993.

Hinden, Robert M. "IP Next Generation." 39 *Communications of the ACM* 61–71 (June 1996).

Hinden, Robert M. "Simple Protocol plus White Paper." RFC 1710. October 1994.

Hirsh, Richard F. *Power Loss: The Origins of Deregulation and Restructuring in the American Electric Utility System.* Cambridge: MIT Press, 1999.

Hirsh, Richard F. *Technology and Transformation in the American Electric Utility Industry.* Cambridge: Cambridge University Press, 1989.

Hoffman, Jeanette. "Governing Technologies and Techniques of Government: Politics on the Net," 1998. Accessed at http://duplox.wzb.eu/final/jeanette.htm.

Hoffman, Jeanette. "Internet Governance: A Regulative Idea in Flux," 2005. Accessed at http://duplox.wzb.eu/people/jeanette/texte/Internet%20Governance%20english%20version.pdf.

Huang, Nen-Fu, Han-Chieh Chao, Reen-Cheng Wang, Whai-En Chen, and Tzu-Fang Sheu. "The IPv6 Deployment and Projects in Taiwan." *Proceedings of the 2003 Symposium on Applications and the Internet Workshops,* IEEE Computer Society, Washington, DC, 2003.

Hughes, Thomas. *American Genesis*. London: Penguin, 1989.

Hughes, Thomas. *Rescuing Prometheus: Four Monumental Projects That Changed the Modern World*. New York: Vintage Books, 1998.

Huitema, Christian. *IPv6: The New Internet Protocol*. Upper Saddle River, NJ: Prentice Hall, 1996.

Huitema, Christian. "The H Ratio for Address Assignment Efficiency." RFC 1715. October 1994.

Huitema, Christian. "Charter of the Internet Architecture Board." RFC 1601. March 1994.

Johnson, David R., Susan P. Crawford, and John G. Palfrey. "The Accountable Net: Peer Production of Governance." 9 *Virginia Journal of Law and Technology* 1–33 (2004).

Kahin, Brian, and Janet Abbate, eds. *Standards Policy for Information Infrastructure*. Cambridge: MIT Press, 1995.

Kempf, James, and Rob Austein. "The Rise of the Middle and the Future of End to End: Reflections on the Evolution of the Internet Architecture." RFC 3724. March 2004.

Knapf, Eric. "Whatever Happened to IPv6?" 31 *Business Communications Review* 14–16 (April 2001).

Kruse, Hans, William Yurcik, and Lawrence Lessig. "The InterNAT: Policy Implications of the Internet Architecture Debate." In *Communications Policy in Transition: The Internet and Beyond*, Benjamin M. Compaine and Shane Greenstein, eds. Cambridge: MIT Press, 2001.

Latour, Bruno. *We Have Never Been Modern*. Cambridge: Harvard University Press, 1993.

Lawton, George. "Is IPv6 Finally Gaining Ground?" *IEEE Computer* 11–15 (August 2001).

Lemley, Mark A., and Lawrence Lessig. "The End of End-to-End: Preserving the Architecture of the Internet in the Broadband Era." UC Berkeley Law and Economics Research Paper 2000–19; Stanford Law and Economics Olin Working Paper 207; UC Berkeley Public Law Research Paper 37 (October 2000).

Lessig, Lawrence. *Code and Other Laws of Cyberspace*. New York: Basic Books, 1999.

Lessig, Lawrence. *Free Culture: How Big Media Uses Technology and the Law to Lock Down Culture and Control Creativity*. New York: Penguin Press, 2004.

Lessig, Lawrence. *The Future of Ideas: The Fate of the Commons in a Connected World*. New York: Random House, 2001.

Libicki, Martin, et al. *Scaffolding the New Web: Standards and Standards Policy for the Digital Economy*. Santa Monica: Rand, 2000.

MacLean, Don. "Herding Schrodinger's Cats: Some Conceptual Tools for Thinking about Internet Governance." *Background Paper for the ITU Workshop on Internet Governance*, Geneva, February 26–27, 2004. Accessed at http://www.itu.int/osg/spu/forum/intgov04/contributions/itu-workshop-feb-04-internet-governance-background.pdf.

Malkin, Gary. "The Tao of IETF, A Guide for New Attendees of the Internet Engineering Task Force." RFC 1718. November 1994.

Mathiason, John, Milton Mueller, Hans Klein, Marc Holitscher, and Lee McKnight. "Internet Governance: the State of Play," Internet Governance Project paper. September 2004. Accessed at http://www.internetgovernance.org/pdf/ig-sop-final.pdf.

McGovern, Michael, and Robert Ullman. "CATNIP: Common Architecture for the Internet." RFC 1707. October 1994.

Morris, John, and Alan Davidson. "Policy Impact Assessments: Considering the Public Interest in Internet Standards Development." 31st Research Conference on Communication, Information and Internet Policy, August 2003. Accessed at http://www.cdt.org/standards.

Mueller, Milton L. *Ruling the Root: Internet Governance and the Taming of Cyberspace*. Cambridge: MIT Press, 2002.

Mueller, Milton L. "Competition in IPv6 Addressing: A Review of the Debate." Concept Paper by the Internet Governance Project, July 5, 2005. Accessed at http://www.internetgovernance.org.

Mueller, Milton L. "The Politics and Issues of Internet Governance." Essay for the Institute on Research and Debate on Governance, April 2007. Accessed at http://www.institut-gouvernance.org/en/analyse/fiche-analyse-265.html.

Mumford, Lewis. *Technics and Civilization*. New York: Harcourt Brace, 1934.

Narten, Thomas, and Richard Draves. "Privacy Extensions for Stateless Address Autoconfiguration in IPv6." RFC 3041. January 2001.

Narten, Thomas. "IETF Statement on IPv4 Exhaustion and IPv6 Deployment." Informational Internet Draft, November 12, 2007.

Ning, Hua. "IPv6 Test-bed Networks and R&D in China." *Proceedings of the 2004 International Symposium on Applications and the Internet Workshops*. Washington, DC: IEEE Computer Society, 2004.

Noble, David. *America By Design: Science, Technology and the Rise of Corporate Capitalism*. New York: Knopf, 1977.

Partridge, Craig, and Frank Kastenholz. "Technical Criteria for Choosing IP the Next Generation (IPng)." RFC 1726. December 1994.

Poovey, Mary. *A History of the Modern Fact: Problems of Knowledge in the Sciences of Wealth and Society.* Chicago: University of Chicago Press, 1998.

Postel, Jon. "DoD Standard Internet Protocol." RFC 760. January 1980.

Postel, Jon. "Assigned Numbers," RFC 739. November 1977.

Postel, Jon, ed. "Internet Protocol: DARPA Internet Program Protocol Specification." RFC 791. September 1981.

Rekhter, Yakov. "An Architecture for IP Address Allocation with CIDR." RFC 1518. September 1993.

Reynolds, Joyce, and Jon Postel. "Assigned Numbers." RFC 870. October 1983.

Reynolds, Joyce, and Jon Postel. "Assigned Numbers." RFC 900. June 1984.

Reynolds, Joyce, and Jon Postel. "Assigned Numbers." RFC 923. October 1984.

Reynolds, Joyce, and Jon Postel. "Assigned Numbers." RFC 943. April 1985.

Reynolds, Joyce, and Jon Postel. "Assigned Numbers." RFC 960. December 1985.

Reynolds, Joyce, and Jon Postel. "Assigned Numbers." RFC 990. November 1986.

Reynolds, Joyce, and Jon Postel. "Assigned Numbers." RFC 1010. May 1987.

RFC Editor, et al. "30 Years of RFCs." RFC 2555. April 1999.

Romano, Sue, et al. "Internet Numbers." RFC 1117. August 1989.

Russell, Andrew. "The American System: A Schumpeterian History of Standardization," *Progress and Freedom Foundation (PFF) Progress on Point Paper 14.4.* (March 2007). Accessed at http://papers.ssrn.com/sol3/papers.cfm?abstract_id=975259.

Saltzer, J., D. P. Reed, and D. D. Clark. "End-to-End Arguments in System Design." 2 *ACM Transactions on Computer Systems* 277–88 (November 1984).

Shapan, Steven. *The Scientific Revolution.* Chicago: University of Chicago Press, 1996.

Shapin, Steven, and Simon Schaffer. *Leviathan and the Air-Pump: Hobbes, Boyle, and the Experimental Life.* Princeton: Princeton University Press, 1985.

Skelton, Ron. "Electric Power Research Institute Comments on IPng." RFC 1673. August 1994.

Slotten, Hugh R. *Radio and Television Regulation: Broadcast Technology in the United States, 1920–1960.* Baltimore: Johns Hopkins University Press, 2000.

Solensky, Frank. *Minutes of the Address Lifetime Expectations Working Group (ALE)*, July 1994. Accessed at ftp://ftp.ietf.cnri.reston.va.us/ietf-onlineproceedings/94jul/ area and.wg.reports/ipng/ale/ale-minutes-94jul.txt.

Sunder, Madhavi. "IP3." University of California, Davis Legal Studies Research Paper 82. 59 *Stanford Law Review* 257–332 (2006).

Taylor, Mark. "A Cellular Industry View of IPng." RFC 1674. August 1994.

Tsirtisis, George, and Pyda Srisuresh. "Network Address Translation—Protocol Translation (NAT-PT)." RFC 2766. February 2000.

Turkle, Sherry. *Life on the Screen: Identity in the Age of the Internet*. New York: Touchstone, 1995.

Vecchi, Mario. "IPng Requirements: A Cable Television Industry Viewpoint." RFC 1686. August 1994.

Vincenti, Walter G. *What Engineers Know and How They Know It: Analytical Studies from Aeronautical History*. Baltimore: Johns Hopkins University Press, 1990.

von Burg, Urs. *The Triumph of Ethernet: Technological Communities and the Battle for the LAN Standard*. Stanford: Stanford University Press, 2001.

Weiser, Mark. "Whatever Happened to the Next-Generation Internet?" 44 *Communications of the ACM* 61–68 (September 2001).

Winner, Langdon. *Autonomous Technology: Technics-out-of-Control as a Theme in Political Thought*. Cambridge: MIT Press, 1977.

Winner, Langdon. "Do Artifacts Have Politics?" 109 *Daedalus* 121–36 (Winter 1980).

Winner, Langdon. "Upon Opening the Black Box and Finding it Empty. Social Constructivism and the Philosophy of Technology." 18 *Science, Technology and Human Values* 362–78 (1993).

Winner, Langdon. *The Whale and the Reactor: A Search for Limits in an Age of High Technology*. Chicago: University of Chicago Press, 1986.

# Index